MACHINE LEARNING

0100010101110100011010000
0110010101101101001000000
0100000101101100011100000
0110000101111001011001010
011010010110111000000110
0000101001001101011000001
0110001101101000011010001
0110111001100101001000001
0100110001100101011000001
0111001001101110011010001
0110111001100111000001101
00001010010101000110100001
0110010100100000010011101
0110010101110111001000001
0100000101001001000001101
00001010

The MIT Press Essential Knowledge Series

A complete list of the titles in this series appears at the back of this book.

MACHINE LEARNING

REVISED AND UPDATED EDITION

ETHEM ALPAYDIN

The MIT Press | Cambridge, Massachusetts | London, England

This book was set in Chaparral Pro by New Best-set Typesetters Ltd. Printed and bound in the United States of America.

Library of Congress Cataloging-in-Publication Data

Names: Alpaydın, Ethem, author.
Title: Machine learning / Ethem Alpaydın.
Description: Revised and updated edition. | Cambridge, Massachusetts : The MIT Press, [2021] | Series: The MIT Press essential knowledge series | Includes bibliographical references and index.
Identifiers: LCCN 2020033697 | ISBN 9780262542524 (paperback)
Subjects: LCSH: Machine learning. | Artificial intelligence.
Classification: LCC Q325.5 .A47 2021 | DDC 006.3/1—dc23
LC record available at https://lccn.loc.gov/2020033697

10 9 8 7 6 5 4 3 2 1

CONTENTS

SERIES FOREWORD

The MIT Press Essential Knowledge series offers accessible, concise, beautifully produced pocket-size books on topics of current interest. Written by leading thinkers, the books in this series deliver expert overviews of subjects that range from the cultural and the historical to the scientific and the technical.

In today's era of instant information gratification, we have ready access to opinions, rationalizations, and superficial descriptions. Much harder to come by is the foundational knowledge that informs a principled understanding of the world. Essential Knowledge books fill that need. Synthesizing specialized subject matter for nonspecialists and engaging critical topics through fundamentals, each of these compact volumes offers readers a point of access to complex ideas.

In November 2019, South Korean Go master Lee Se-dol announced that, after a career of twenty-four years, he was retiring from professional Go competitions. In 2016, he had played a five-game series against a computer program named AlphaGo, which he lost 1 to 4. Since then, later versions of AlphaGo had gotten even better, so much so that when announcing his retirement, Se-dol said that, "With the debut of AI in Go games, I've realized that I'm not at the top even if I become the number one through frantic efforts. Even if I become the number one, there is an entity that cannot be defeated."[1]

The ancient strategy game Go had long been believed to be beyond the capability of AI. In 1997, when the chess-playing program Deep Blue defeated the reigning world champion Garry Kasparov, researchers believed that it would take another generation for the same to happen with Go. While the chess board is 8 × 8, the Go board is 19 × 19—a larger board meaning so many more possible positionings of the pieces on the board exist, hence so many more different ways of playing the game that a game-playing computer program should be able to handle.

The crucial difference between Deep Blue and AlphaGo was the shift from programming to learning. Whereas Deep Blue was programmed by human experts to play as well as possible, AlphaGo learned to play well by playing many games and updating itself using this experience, favoring moves and strategies that led to winning the game and penalizing those that led to losses.

How this is done is the topic of this book, and as we will see, game playing is only one of the many domains where we have witnessed this unforeseen sudden jump in ability through learning. In the last two decades, using systems that learn, we have seen drastic improvements in accuracy in various applications that have since been successfully commercialized. We now have programs that can recognize people from their faces, understand spoken speech, recommend a movie, translate text from one language to another, and drive a car—all of which have been made possible by machine learning.

Once, it used to be the programmer who had to come up with a way to solve the problem; the sequence of operations that needs to be carried out is named an algorithm. The algorithm is then coded as a program using a programming language, and the program is executed on a computer. In a learning program, on the other hand, the programmer specifies how the data (collected through experience) is used to update the program so as to improve performance; it is the data that determines the final form of the program.

In a programmed system, the programmer knows at the time of writing the program how the system is going to behave in any situation. The program has no intelligence by itself; it is just a machine that is hardwired to duplicate the intelligence of the programmer. It just does what the programmer would do; its only advantage may be its speed; it is no more than a calculator.

With learning, however, how a system will act in a situation is the result of the interaction between the learning program and the data, and as we will see, the final system very much depends on the quantity and quality of the data (i.e., how well the data covers all possible scenarios). In such a case, how a trained program will act cannot be foreseen by the programmer at the time of writing the program, and as such it can be said that a program that has learned from data has acquired intelligence beyond that of the programmer.

In retrospect, it is not surprising that the learning program AlphaGo defeated Lee Se-dol. AlphaGo played (and learned from) many more games than any human being can play in a lifetime. Likewise, a doctor gains experience from their own patients only; a learning medical diagnosis system can be trained with the collection of patients of thousands of doctors. Similarly, a car that learns to drive itself can be trained with many more and much more varied scenarios than even the most experienced human driver can encounter in a lifetime. That is the advantage of collecting big data and analyzing it to infer knowledge.

Of course, learning from data is not new; it is at the heart of science. In the past, scientists like Galileo and

Kepler designed experiments to make observations and collected data; they then came up with laws that explain those data. In medicine, cures for many diseases were found by collecting information from patients and analyzing them for commonalities and differences. But we are now at a point where we want to automate this process of going from data to knowledge, because now we have much more data and many more application domains.

Since the advent of computers in the middle of the last century, our lives have become increasingly computerized and digital. Computers are no longer just the numeric calculators they once were. Databases and digital media have taken the place of printing on paper as the main medium of information storage, and digital communication over computer networks supplanted the post as the main mode of information transfer. First with the personal computer with its easy-to-use graphical interface, and then with the phone and other smart devices, the computer has become a ubiquitous device, a household appliance just like the TV or the microwave. Currently, all sorts of information, not only numbers and text but also image, video, audio, and so on, are stored, processed, and—thanks to online connectivity—transferred digitally. All this digital processing results in a lot of data, and it is this surge of data that is mainly responsible for triggering the widespread interest in data analysis and machine learning.

For many applications—from vision to speech, from translation to robotics—we were not able to devise very good algorithms despite decades of research beginning in the 1950s. But for all these tasks, it is easy to collect data, and now the idea is to learn the algorithms for these automatically from data, replacing programmers with learning programs. This is the niche of machine learning, and it is not only that the data has gotten bigger in these last two decades, but also that the theory of machine learning to process that data to turn it into knowledge has advanced significantly.

Once, if you were smart, you invented a new algorithm; now, if you are smart, you find a new source of data, possibly first by writing the app to collect it. In the past, computer science advanced one algorithm at a time; now information technology advances one app at a time.

Today, in different types of business, from retail and finance to manufacturing, as our systems are computerized, more data is continuously generated and collected. This is also true in various fields of science, from astronomy to biology. In our everyday lives too, as digital technology increasingly infiltrates our daily existence, as our digital footprint deepens, not only as consumers and users but also through social media, an increasingly larger part of our lives is recorded and becomes data. Whatever its source—business, scientific, or personal—data that just lies dormant passively is not of any use, and smart people have been finding new ways to make use of that data and turn it into a useful product or service. In this transformation, machine learning is playing a more significant role.

Our belief is that behind all this seemingly complex and voluminous data, there lies a simple explanation. That although the data is big, it can be explained in terms of a relatively simple model with a small number of hidden factors and their interaction. Think about millions of customers who buy thousands of products online or from their local supermarket every day. This implies a very large database of transactions; but what saves us and works to our advantage is that there is a pattern to this data. People do not shop at random. A person throwing a party buys a certain subset of products, and a person who has a baby at home buys a different subset; there are hidden factors that explain customer behavior. It is this inference of a hidden model—namely, the underlying factors and their interaction—from the observed data that is at the core of machine learning.

Machine learning is not just the commercial application of methods to extract information from data; learning is also a requisite of intelligence. An intelligent system should be able to adapt to its environment; it should learn not to repeat its mistakes but to repeat its successes. Previously, researchers used to believe that for artificial intelligence to become reality, we needed a new paradigm, a new type of thinking, a new model of computation, or a whole new set of algorithms. Taking into account the recent successes in machine learning in various domains, we can now claim that what we need is not a new set of specific algorithms but a lot of example data and sufficient computing power to run the learning methods on that much data, bootstrapping the necessary algorithms from data.

It appears that tasks such as machine translation and planning can be solved with learning algorithms that are relatively simple but trained on large amounts of example data. Recent successes with "deep learning" (e.g., the AlphaGo program) support this claim. Intelligence seems not to originate from some outlandish formula, but rather from the patient, almost brute force use of simple, straightforward algorithms.

As technology develops and we get faster computers and more data, learning algorithms can be expected to generate a slightly higher level of intelligence, which will find use in a new set of slightly smarter devices and software. It will not be surprising if this type of learned intelligence reaches the level of human intelligence some time before this century is over.

I believe that this is the perspective from which we can assess the meaning of Lee Se-dol's retirement. Some people may find him unnecessarily touchy; athletes did not stop running marathons when cars were invented. Playing games such as chess or Go will always be a test of a person's ability to focus and strategize. People are not going to stop playing these games just because a computer can beat them; on the contrary, game-playing computer programs have the potential to teach them how to play those games better.

We can see Lee Se-dol's resignation as another step toward humanity no longer being either the standard by which intelligence is measured or the maximum intelligence that can be attained. What we have been encountering in the last half century as digital technology has advanced one milestone at a time (and I believe AlphaGo is one such milestone) is similar to the Copernican revolution where the earth was demoted from being the center of the heavens to simply one of the planets in the solar system.

We are systematically noticing that what we have with our brains is only one way of being intelligent, and not necessarily the best one at that. It is amusing that in 2016, when Lee Se-dol lost to AlphaGo, this made headlines all around the world because this was the first time a computer program defeated the best human player; when he retired in 2019, the news article referred to him as "the only human to defeat the AI Go player AlphaGo."

The aim of this book is to give the reader an overall idea about what machine learning is, the basics of some important learning algorithms, and a set of example applications.

The book is intended for a general readership, and only the essentials of the learning methods are discussed without any mathematical or programming details. The book does not cover any of the machine learning applications in much detail either; a number of examples are discussed just enough to give the fundamentals without going into the particulars.

For more information on machine learning algorithms and applications, the reader can refer to my textbook on the topic, on which this book is based: Ethem Alpaydın, *Introduction to Machine Learning*, 4th ed. (Cambridge, MA: MIT Press, 2020).

The content is organized as follows:

We start by briefly discussing the evolution of computer science and its applications in chapter 1. This is necessary to place in context the current state of affairs that created the interest in machine learning—namely, how the digital technology advanced from number-crunching mainframes to desktop personal computers and later on to smart devices that are online and mobile. This chapter shows how we have ended up with so much data.

Chapter 2 introduces the basics of machine learning and discusses how it relates to model fitting and statistics on some simple applications.

Most machine learning algorithms are supervised, and in chapter 3, we discuss how such algorithms are used for pattern recognition, such as faces and speech.

Chapter 4 covers artificial neural networks inspired from the human brain, how they can learn, and how "deep," multilayered networks can learn hierarchies at different levels of abstractions.

Another type of machine learning is unsupervised, where the aim is to learn associations between instances. In chapter 5, we consider customer segmentation and learning recommendations as popular applications.

Chapter 6 is on reinforcement learning where an autonomous agent—for example, a self-driving car—learns to take actions in an environment to maximize reward and minimize penalty.

As any new technology, since we have started using machine learning in the real world, we have started to face its concomitant challenges and risks. So this new edition has a new chapter, chapter 7, for topics that have become increasingly important since 2016 when the first edition appeared. These include ethical and legal implications of automated decision making, concerns over data privacy and security, possible biases in training data, the need for explainability, and others.

Chapter 8 concludes by discussing some future directions and the newly proposed field of "data science" that also encompasses high-performance cloud computing.

This book aims to give a quick introduction to what is being done today in machine learning, and my hope is to trigger the reader's interest in thinking about what can be done in the future. Machine learning is certainly one of the most exciting scientific fields today, advancing technology in various domains, and it has already generated a set of impressive applications affecting all walks of life. I have enjoyed very much writing this book; I hope you will enjoy reading it too!

I am grateful to the anonymous reviewers for their constructive comments and suggestions. As always, it has been a pleasure to work with the MIT Press and I would like to thank the anonymous reviewers, Kathleen Caruso, and Marie Lufkin Lee for all their support.

WHY WE ARE INTERESTED IN MACHINE LEARNING

The Power of the Digital

Some of the biggest transformations in our lives in the last half century are due to computing and digital technology. The tools, devices, and services we had invented and developed in the centuries before have been increasingly replaced by their computerized "e-" versions, and we in turn have been continuously adapting to this new digital environment.

This transformation has been very fast: once upon a time—fifty years ago is mythical past in the digital realm where things happen at the speed of light—computers were expensive and only very large organizations, such as governments, big companies, universities, research centers, and so on, could afford them. At that time, only they had problems difficult enough to justify the high cost of procuring and maintaining a computer. Computer "centers," in separate floors or buildings, housed those power-hungry behemoths, and inside large halls, magnetic tapes whirred, cards were punched, numbers were crunched, and bugs were real bugs.

As computers became cheaper, they became available to a larger selection of the population and in parallel, their application areas widened. In the beginning, computers were nothing but calculators—they added, subtracted, multiplied, and divided numbers to get new numbers. Probably the major driving force of the computing technology is the realization that every piece of information can be represented as numbers. This in turn implies that the computer, which until then was used to process numbers, can be used to process all types of information.

To be more precise, a computer represents every number as a particular sequence of binary digits (bits) of 0 or 1, and such bit sequences can also represent other types of information. For example, "101100" can be used to represent the decimal 44 *and* is also the code for the comma; likewise, "1000001" is both 65 and the uppercase letter 'A'.[1] Depending on the context, the computer program manipulates the sequence according to one of the interpretations.

Actually, such bit sequences can stand for not only numbers and text, but also other types of information—for example, colors in a photo or tones in a song. Even computer programs are sequences of bits. Furthermore, and just as important, operations associated with these types of information, such as making an image brighter or finding a face in a photo, can be converted to commands that manipulate bit sequences.

Computers Store Data

The power of the computer lies in the fact that every piece of information can be represented digitally—that is, as a sequence of bits—and every type of information processing can be written down as computer instructions that manipulate such digital representations.

One consequence of this emerged in the 1960s with *databases*, which are specialized computer programs that store and manipulate *data*, or digitally represented information. Peripheral devices, such as tapes or disks, store bits magnetically, and hence their contents are not erased when the computer is switched off.

With databases, computers have moved beyond processing and have become repositories of information using the digital representation. In time, the digital medium has become so fast, cheap, and reliable that it has supplanted printing on paper as humanity's main means of information storage.

After the invention of the microprocessor and parallel advances in miniaturization and decreasing costs, personal computers became increasingly available starting in the early 1980s. The *personal computer* has made computing accessible to small businesses, but most important the personal computer was small and cheap enough to be a household appliance. You did not need to be a large company; the computer could help with your life too. The personal computer confirmed that everyone had tasks that are computer-worthy, and the growth of applications followed this era of democratization of digital technology.

Graphical user interfaces and the mouse made the personal computer easy to use. We do not need to learn programming, or memorize commands with a difficult syntax, to be able to use the computer. The screen is a digital simulacrum of our work environment displaying a virtual desktop, with files, icons, and even a trash can, and the mouse is our virtual hand that picks them up to read or edit them.

The software for the personal computer in parallel has moved from commercial to personal applications by manipulating more types of data and making more of our lives digital. We have a word processor for our letters and other personal documents, a spreadsheet for our household calculations, and software for our hobbies such as music or photography; we can even play games if we want to! Computing has become everyday and fun.

The personal computer with its pleasant and inviting user interface coupled with a palette of everyday applications was a big step in the rapprochement between people and computers, our life as we used to know it, and the digital world. Computers were programmed to fit our lives a little better, and we have adapted a little to accommodate them. In time, using a computer has become a basic skill, like driving.

The personal computer was the first step in making computers accessible to the masses; it made digital technology a larger part of our lives and, most important for our story in this book, allowed more of our lives to be recorded digitally. As such, it was a significant stepping-stone in this process of converting our lives to data, data that we can then analyze and learn from.

Computers Exchange Data

The next major development in computing was in connectivity. Though hooking up computers by data links to exchange information had been done before, commercial systems started to become widespread in 1990s to connect personal computers to each other or to central servers over phone or dedicated lines.

The computer network implies that a computer is no longer isolated but can exchange data with a computer far away. A user is no longer limited to their own data on their own computer but can access data elsewhere, and if they want, they can make their data available to other users.

The development of computer networks very quickly culminated in the Internet, which is *the* computer network that covers the whole globe. The Internet made it possible for anyone in the world who has access to a computer to send information, such as an email, to anyone else. And because all our data and devices are already digital, the information we can share is more than just text and numbers; we can send images, videos, music, speech, anything.

With computer networks, digitally represented information can be sent at the speed of light to anyone, anywhere. The computer is no longer just a machine where data is stored and processed, but it has also become a means to transfer and share information. Connectivity increased so quickly and digital communication has become so cheap, fast, and reliable that digital transfer has supplanted mail as the main technology for information transfer.

Anyone who is "online" can make their own data on their own computer available over the network to anyone else, which is how the World Wide Web was born. People can "surf" the Web and browse this shared information. Very quickly, secure protocols have been implemented to share confidential information, thereby permitting commercial transactions over the Web, such as online shopping or banking. Online connectivity has further increased the infiltration of digital technology. When we get an online service by using the "www." portal of the service provider, our computer turns into the digital version of the shop, the bank, the library, or the university; this, in turn, created more data.

Mobile Computing

Every decade we have been seeing computers getting smaller, and with advances in battery technology,[2] in the mid-1990s, portable—laptop or notebook—computers that can also run on batteries started to become widespread; this started the new era of *mobile computing*. Cellular phones also started to become popular around the same time, and roughly around 2005, these two technologies merged in the smartphone.

A *smartphone* is a phone that is also a computer. In time, the smartphone became smarter—more a computer and less a phone—so much so that today, the phone is only one of many apps on a smartphone, and a rarely used one at that. The traditional phone was an acoustic device: you talked into it, and you heard the person on the other end talking. The smartphone today is more of a visual device; it has a large screen and we spend more time looking at this screen or tapping its touch-sensitive surface than talking.

A smartphone is a computer that is always online[3] and it allows its user to access the Internet for all types of information while mobile. It therefore extends our connectivity in that it permits us greater access—for example, while traveling—to data on other computers, as well as making us, and our data, accessible to others.

What makes a smartphone special is that it is also a mobile sensing device and because it is always on our person, it continuously records information about us, most notably our position, and can make this data available. The smartphone is a mobile sensor that makes us detectable, traceable, recordable.

This increased mobility of the computer is new. Once the computer was big and at a "computer center"; it stayed fixed, and we needed to walk to it. We sat in front of a terminal to use the computer—it was called a "terminal" because the computer ended there. Then a smaller computer came to our department, and later a smaller one sat on our desk in our office or in our house, and then an even smaller one was on our lap, and now the computer is in our pocket and with us all the time.

Once there were very few computers, possibly one computer per thousands of people—for example, one per company or campus. This computer-per-person ratio increased very quickly, and the personal computer aimed to have one computer for every person. Today we have many computers per person. Now all our devices are also computers or have computers in them. Your phone is also a computer, your TV is also a computer, your car has many computers inside it for different functions, and your music player is a specialized computer as is your camera or your watch. The *smart device* is a computer that does the digital version of whatever it did originally.

Ubiquitous computing is a term that is becoming increasingly popular; it means using computers without knowing that you are using one. It means using a lot of computers for all sorts of purposes all the time without explicitly calling them computers. The digital version has its usual advantages, such as speed, accuracy, and easy adaptability. But another advantage that is most relevant to our discussion is that the digital version of the device now has all its data digitally. And furthermore, if it is online, it can talk to other online computers and make its data available almost instantaneously. We call them "smart objects" or just "things" and talk about the *Internet of Things*.

Social Data

A few thousands of years ago, you needed to be a god or goddess if you wanted to be painted, be sculpted, or have your story remembered and told. A thousand years ago you needed to be a king or queen, and a few centuries ago you needed to be a rich merchant, or in the household of one. Now anybody, even a soup can, can be painted. A similar democratization has also taken place in computing and data. Once only large organizations and businesses had tasks worthy of a computer and hence only they had data; starting with the personal computer, people and even objects became generators of data.

With most communication now being done using computers (including smartphones) over the Internet, a recent source of data is *social media*, where our social interactions have become digital; these now constitute another type of data that can be collected, stored, and analyzed. Social media replaces discussions in the agora, piazza, market, coffeehouse, and pub, or at the gathering by the spring, the well, and the water cooler.

With social media, each of us is now a celebrity whose life is worth following, and we are our own paparazzi. We are no longer allotted only fifteen minutes of fame, but every time we are online we are famous. The social media allows us to write our digital autobiography as we are living it. In the old times, books and newspapers were expensive and hence scarce; we could keep track of and tell the story of only important lives. Now data is cheap, and we are all kings and queens of our little online fiefdoms. Digitally savvy people of today who are continually posting on the social media as they wander from place to place are modern day versions of Odysseus, composing, and at the same time broadcasting in real time, their own digitized epics, with their own thrills and dangers.

All That Data: The Dataquake

The data generated by all our computerized machines and services was once a by-product of digital technology, and computer scientists have done significant amount research on how to store and manipulate large amounts of data in databases most efficiently. Then, we stored data because we had to; more data meant costlier storage and slower access. Sometime in the last two decades, all this data became a resource; now, more data is a blessing.

Think, for example, of a supermarket chain that sells thousands of goods to millions of customers every day, either at one of the numerous brick-and-mortar stores all over a country or through a virtual store over the Web. The point-of-sales terminals are digital and record the details of each transaction: data, customer id (through some loyalty program), goods bought and at what price, total money spent, and so forth. The stores are connected online, and the data from all the terminals in all the stores can be instantaneously collected in a central database. This amounts to a lot of (and very up-to-date) data every day.

Especially in the last twenty years or so, people have increasingly started to ask themselves what they can do with all this data. With this question the whole direction of computing is reversed. Before, data was what the programs processed and spit out—data was passive. With machine learning, data starts to drive the operation; it is not the programmers anymore but the data itself that defines what to do next.

One thing that a supermarket chain is always eager to know is which customer is likely to buy which product. With this knowledge, stores can be stocked more efficiently, which will increase sales and profit. It will also increase customer satisfaction because customers will be able to find the products that they need quicker and cheaper.

This task is not evident. We do not know exactly which people are likely to buy this ice cream flavor or the next book of this author, to see this new movie, or to visit this city. Customer behavior changes in time and depends on geographic location.

But there is hope, because we know that customer behavior is not completely random. People do not go to supermarkets and buy things at random. When they buy beer, they buy chips; they buy ice cream in summer and spices for Glühwein in winter. There are certain patterns in customer behavior, and that is where data comes into play.

Though we do not know the customer behavior patterns themselves, we expect to see them occurring in the collected data. So if we can find such patterns in past data,

With machine learning, data starts to drive the operation; it is not the programmers anymore but the data itself that defines what to do next.

assuming that the future, at least the near future, will not be much different from the past when that data was collected, we could expect them to continue to hold, and we can make correct predictions based on them.

We may not be able to identify the process completely, but we believe we can construct *a good and useful approximation*. That approximation may not explain everything but may still be able to account for some part of the data. We believe that though identifying the complete process may not be possible, we can still detect some patterns. We can use those patterns to predict; they may also help us understand the process.

This is called *data mining*. The analogy is that a large volume of earth and raw material is extracted from the mine, which when processed leads to a small amount of very precious material. Similarly, in data mining, a large volume of data is processed to construct a simple model with valuable use, for example, one with high predictive accuracy.

Data mining is one type of machine learning. We do not know the rules (of customer behavior), so we cannot write the program, but the machine—that is, the computer—"learns" by extracting such rules from (customer transaction) data.

Many applications exist where we do not know the rules but have a lot of data. As we discussed before, the fact that computers and digital technology have penetrated so deep into our everyday lives implies that now there are large amounts of data in all sorts of domains ready for mining.

Learning models are also used in pattern recognition, for example, in recognizing images captured by a camera or recognizing speech captured by a microphone. These days, we have different types of sensors used for different type of applications, from human activity recognition using a smartphone to driving assistance systems in cars.

Another data source is science. As we build better sensors, we detect more—that is, we get more data—in astronomy, biology, physics, and so on, and we use learning algorithms to make sense of the bigger and bigger data. The Internet itself is one huge data repository, and we need smart algorithms to help us find what we are looking for. One important characteristic of data we have today is that it comes from different modalities—it is multimedia. We have text, we have images or video, we have sound clips, and so on, all somehow related to the same object or event we are interested in, and a major challenge in machine learning today is to combine information coming from these different sources. For example, in consumer data analysis, in addition to past transactions, we also have Web logs—namely, the Web pages that a user has visited recently—and these logs may be quite informative.

With all types of smart machines continuously helping us in our daily lives, we have all become producers of data. Every time we buy a product, every time we rent a movie, visit a Web page, or post on the social media, even when we just walk or drive around, we are generating data. And that data is valuable for someone who is interested in collecting and analyzing it, because we are also consumers of data. We want to have products and services specialized for us. We want our needs to be understood and our interests to be predicted. The customer is not only always right, but also interesting and worth tracking.

Learning versus Programming

To solve a problem on a computer, we need an algorithm. An algorithm is a sequence of instructions that are carried out to transform the input to the output. For example, we have an algorithm for calculating payroll: The input is the work-related information of an employee, such as a timesheet, and personal information such as marital status, and the output is their salary.

An algorithm is similar to a recipe for a dish. Preparing any dish requires basic actions such as peeling, slicing, frying, and so on. The recipe for a dish defines which of these actions should be carried out on which ingredient and in which order. Any person who can do these basic actions can prepare a dish just by following the recipe. This is what we have in computer programming, where the central processing unit (CPU) of the computer has a set of basic instructions and the algorithm defines which instructions should be carried out on which input and in which order. A software library is just like a cookbook.

In the past decades we have devised algorithms for many tasks, and that is why computers and digital technology are so widely used now. But for some problems, we do not yet have an algorithm. Predicting customer behavior is one; another is differentiating spam emails from legitimate ones. We know what the input is: an email document that in the simplest case is a text message. We know what the output should be: a yes/no output indicating whether the message is spam or not. But we do not know how to transform the input to the output. What is considered spam changes over time and from individual to individual.

What we lack in knowledge, we make up for in data. We can easily compile thousands of messages, some of which we know to be spam and some of which are not, and what we want is to "learn" what constitutes spam from this sample. For example, after analyzing example data, we may notice that words like "offer" and "opportunity" or symbols like "$" or "!" appear much more frequently in spam e-mails than they appear in ordinary e-mails, so our spam filter increases the probability that a given e-mail is spam if it sees any one of these in a new email.

In learning, we would like the computer (the machine) to extract automatically the algorithm for the task that underlies the data. There is no need to learn to calculate payroll, we already know how to do it, but there are many applications for which we do not have an algorithm but can collect lots of data.

Artificial Intelligence

Machine learning is not just a database or programming problem; it is also a requirement for artificial intelligence. A system that is in a changing environment should have the ability to learn; if it keeps on making the same mistakes over and over, we will hardly call it intelligent. Learning is smart from an engineering point of view as well because if the system can detect and adapt to changes, the system designer need not foresee and provide solutions for all possible situations.

For us, the system designer was evolution, and our body shape as well as our built-in instincts and reflexes have evolved over millions of years. We also learn to change our behavior during our lifetime. This helps us cope with changes in the environment that cannot be predicted by evolution. Organisms that have a short life in a well-defined environment may have all their behavior built-in, but instead of hardwiring into us all sorts of behavior for any circumstance that we might encounter in our life, evolution gave us a large brain and a mechanism to learn such

Machine learning is not just a database or programming problem; it is also a requirement for artificial intelligence. A system that is in a changing environment should have the ability to learn.

that we could update ourselves with experience and adapt to different environments. That is why human beings have survived and prospered in different parts of the globe in very different climates and conditions. When we learn the best strategy in a certain situation that knowledge is stored in our brain, and when the situation arises again—when we recognize ("cognize" means to know) the situation—we recall the suitable strategy and act accordingly.

Each of us, actually every animal, is a data scientist. We collect information about our environment using our sensors, and then we process the data to devise rules of behavior to control our actions in different circumstances to minimize pain and/or maximize pleasure. We have memory to store those rules in our brains, and then we recall and use them when needed. Learning is lifelong; we forget rules when they no longer apply or revise them when the environment changes.

Learning has its limits though; there may be things that we can never learn with the limited capacity of our brains, just like we can never "learn" to grow a third arm, or an eye in the back of our head—something that would require a change in our genetic makeup. Roughly speaking, genetics defines the hardware that slowly adapts over thousands of generations through mutation, whereas learning defines the adaptation of the software running on (and being constrained by) that hardware during an individual's lifetime.

Artificial intelligence takes inspiration from the brain. There are cognitive scientists and neuroscientists whose aim is to understand the functioning of the brain, and toward this aim, they build models of neural networks and make simulation studies. But artificial intelligence is a part of computer science and our aim is to build useful systems, as in any domain of engineering. So, though the brain inspires us, ultimately, we do not care much about the biological plausibility of the algorithms we develop.

We are interested in the brain because we believe that it may help us build better computer systems. The brain is an information-processing device that has some incredible abilities and surpasses current engineering products in many domains—for example, vision, speech recognition, and learning, to name three. These applications have evident economic utility if implemented on machines. If we can understand how the brain performs these functions, we can define solutions to these tasks as formal algorithms and implement them on computers.

Computers were once called "electronic brains," but computers and brains are different. Whereas a computer generally has one or few processors, the brain is composed of a very large number of processing units, namely, neurons, operating in parallel. Though the details are not completely known, the processing units are believed to be much simpler and slower than a typical processor in a computer. What also makes the brain different, and is believed to provide its computational power, is its large connectivity. Neurons in the brain have connections, called synapses, to tens of thousands of other neurons, and they all operate in parallel. In a computer, the processor is active, and the memory is separate and passive, but it is believed that in the brain both processing and memory are distributed together over the network; processing is done by the neurons and memory occurs in the synapses between the neurons.

Understanding the Brain

According to Marr (1982), understanding an information processing system works at three *levels of analysis*:

1. *Computational theory* corresponds to the goal of computation and an abstract definition of the task.

2. *Representation and algorithm* define how the input and the output are represented, and about the specification of the algorithm for the transformation from the input to the output.

3. *Hardware implementation* is the actual physical realization of the system.

The basic idea in these levels of analysis is that for the same computational theory, there may be multiple representations and algorithms manipulating symbols in that representation. Similarly, for any given representation and algorithm, there may be multiple hardware implementations. For any theory, we can use one of various algorithms, and the same algorithm can have different hardware implementations.

Let us take an example: '6', 'VI', and '110' are three different representations of the number six; respectively, they are the Arabic, Roman, and binary representations. There is a different algorithm for addition depending on the representation used. Digital computers use the binary representation and have circuitry to add in this representation, which is one particular hardware implementation. Numbers are represented differently, and addition corresponds to a different set of instructions on an abacus, which is another physical implementation. When we add two numbers "in our head," we use another representation and an algorithm suitable to that representation, which is implemented by the neurons. But all these different physical realizations—namely, us, abacus, digital computer—implement the same computational theory: addition.

In engineering we go from top to bottom. For example, in software engineering, first we decide on the requirements, then we go down one step and devise an algorithm suitable for the task and the suitable data structures to store the necessary information in computer memory,[4] and finally we go down another step and write the algorithm in a programming language to be executed on a particular computer.

Sometimes it is necessary to go in the opposite direction from bottom to top; this is called *reverse engineering*. For example, in World War II, the German military used a machine called the Enigma to encrypt communications. Intensive work by the Allies resulted in the discovery of its internal mechanisms, and this allowed decryption.

Another example, though this is not information processing, is the difference between natural and artificial flying machines. Humanity had always dreamed of flying, and our early attempts to copy birds by putting on big wings did not work; our arms and shoulders are not strong enough to flap wings big enough to carry our weight—that was an attempt to copy at too low a level. But once science advanced and we discovered the rules of aerodynamics, that is, once we were able to go up to the level of theory, we were able to devise another implementation for the same theory, and with the means we had, we invented propellers, and later on, jet engines. A sparrow flaps its wings; an airplane does not flap its wings but uses jet engines. The sparrow and the airplane are two hardware implementations built for different purposes, satisfying different constraints. But they both obey the same theory, which is aerodynamics.

In artificial intelligence, we want to do the same for intelligence. We can say that the brain is one hardware implementation for intelligence. If from this particular implementation we can do reverse engineering and extract the representation and the algorithm used, and if from that in turn we can get the computational theory, we can then use another representation and algorithm, and in turn a hardware implementation more suited to the means and constraints we have.

In chapter 4 we will discuss artificial neural networks that are composed of interconnected processing units and how such networks can learn—this is the representation and algorithm level. In time, when we discover the computational theory of intelligence, we may discover that neurons and synapses are implementation details, just as we have realized that feathers are for flying.

Pattern Recognition

In computer science, we have tried to solve many tasks using manually specified rules and algorithms. Decades of work have led to very limited success. Some of these tasks relate to artificial intelligence in that they are believed to require intelligence. The current approach, in which we have seen tremendous progress recently, is to use machine learning from data.

Let us take the example of recognizing faces: this is a task that we do effortlessly; every day we recognize family members and friends by looking at their faces or from their photographs, despite differences in pose, lighting, hairstyle, and so forth. Face perception is important for us because we want to tell friend from foe. It was important for our survival not only for identification, but also because the face is the dashboard of our internal state. Feelings such as happiness, anger, surprise, and shame can be read from our face, and we have evolved both to display such states as well as to detect it in others.

Though we do such recognition easily, we do it unconsciously and are unable to explain how we do it. Because we are not able to explain our expertise, we cannot write the corresponding computer program.

By analyzing different face images of a person, a learning program captures the pattern specific to that person and then checks for that pattern in a given image. This is one example of pattern recognition.

The reason we can do this is because we know that a face image, just like any natural image, is not just a random collection of pixels (a random image would be like a snowy TV). A face has structure. It is symmetric. The eyes, the nose, and the mouth are located in certain places on the face. Each person's face is a pattern composed of a particular combination of these. When the illumination or pose changes, when we grow our hair or put on glasses, or when we age, certain parts of the face image change but some parts do not. The learning algorithm finds those unchanging discriminatory features and the way they are combined to define a particular person's face by going over a number of images of that person.

What We Talk about When We Talk about Learning

In machine learning, the aim is to have a computer program that learns. Learning means getting better through experience. "Better" implies a performance criterion that is optimized. "Experience" implies data collected in the past—for example, in past trials. From a programming point of view, "getting better" is implemented as the modification of the decision-making program so that in time, as it sees more data, its output leads to higher performance according to the criterion that is to be optimized.

A learning program is different from an ordinary computer program in that it is a *general template with modifiable parameters*, and by assigning different values to these parameters the program can do all sorts of different things. The learning algorithm *adjusts the parameters* of the template—which we call a model—by *optimizing a performance criterion defined on the data*.

For example, for a game-playing program, the parameters are adjusted as we play against an opponent so that the ratio of our wins to losses increase. For a face recognizer, the parameters are adjusted so that we get the highest prediction accuracy on a set of training images of a person. The learning process is generally repetitive and incremental. The learning program sees examples (games or faces) one after the other, and the parameters are updated slightly at each example, so that the performance improves gradually in time. After all, this is what learning is: as we learn a task, we get better at it, be it tennis, geometry, or a foreign language.

In chapter 2, we will cover in more detail what the template is (actually as we will see, we have different templates depending on the type of the task) and the different learning algorithms that adjust the parameters so as to get the best performance.

In building a learner, there are a number of important considerations:

First, we should keep in mind that just because we have a lot of data, it does not mean that there are underlying rules that can be learned. We should make sure that there are dependencies in the underlying process and that the collected data provides enough information for them to be learned with acceptable accuracy.

Let's say we have a phone book containing people's names and their phone numbers. It does not make sense to believe that there is an overall relationship between names and phone numbers; in such a case, we can do no

The learning program sees examples one after the other, and the parameters are updated slightly at each example, so that the performance improves gradually. This is what learning is: as we learn a task, we get better at it.

better than just storing the known name–phone number pairs in a database. And furthermore, there can be no generalization to new instances; because we cannot infer a general rule, we cannot make a prediction for the phone number of a new name.

Second, the learning algorithm itself should be efficient, because generally we have a lot of data and we want learning to be as fast as possible, using computation and memory effectively. In many applications, the underlying characteristics of the problem may change in time; in such a case, previously collected data becomes obsolete and the need arises to continuously and efficiently update the trained model with new data.

Third, once a learner has been built and we start using it for prediction, it should be efficient in terms of memory and computation as well. In certain applications, the efficiency of the final model may be as important as its predictive accuracy. For the case of a self-driving car for example, all the necessary computation for recognition, decision making, and action should be done fast enough so that the car can go at a reasonable speed.

History

Going from particular examples to general concepts is called *induction*. Over the course of our lives, we see many trees at different times in different places, all slightly different from the other trees in some respects; yet, at the same time, they also all have something in common, and it is this set of common properties that defines our general definition of "treeness." That general concept is stored in our mind so that when we see a new object, we can say whether it is a tree or not depending on how well that object matches our learned definition of "treeness."

Almost all of science is fitting general models to data. Scientists—such as Galileo, Newton, and Mendel—designed experiments, made observations, and collected data. They then tried to extract knowledge by devising theories, that is, by building models to explain the data they observed.[5] They then used these theories to make predictions and if they didn't work, they collected more data and revised the theories. This process of data collection and theory/model building continued until they got models that had enough explanation power.

We are now at a point where this type of data analysis can no longer be done manually, because people who can do such analysis are rare; furthermore, the amount of data is huge and manual analysis is not possible. There is thus a growing interest in computer programs that can analyze data and extract information automatically from them—in other words, learn.

The methods that we discuss have their origins in different scientific domains. It was not uncommon that sometimes the same or very similar algorithms were independently invented in different fields following different historical paths.

The main theory underlying machine learning comes from statistics, where going from particular observations, called the *sample*, to general descriptions of the *population*, is called *inference* and learning is called *estimation*. Classification is called *discriminant analysis* in statistics. Statisticians used to work on small samples and, being mathematicians, mostly worked on simple models that could be analyzed mathematically. In engineering, classification is called pattern recognition and the approach is more empirical.

In computer science, as part of work on artificial intelligence, research was done on learning algorithms; a parallel but almost independent line of study was called *knowledge discovery in databases.* In electrical engineering, research in signal processing resulted in adaptive image processing and speech recognition programs.

In the mid-1980s, a huge explosion of interest in artificial neural network models from various disciplines took place. These disciplines included physics, statistics, psychology, cognitive science, neuroscience, and linguistics, not to mention computer science, electrical engineering, and adaptive control. Perhaps the most important contribution of research on artificial neural networks is this synergy that bridged various disciplines, especially statistics and computer science. The fact that neural network research, which later led to the field of machine learning, started in the 1980s is not accidental. At that time, with advances in VLSI (very large-scale integration) technology, we gained the capacity to build parallel hardware containing thousands of processors, and artificial neural networks was of interest as a possible theory to distribute computation over a large number of processing units, all running in parallel. Furthermore, because they could learn from data, they would not need programming.

As we will discuss in chapter 4, a neural network is composed of layers of processing units, each checking for a particular condition in the input and with successive layers checking for more abstract conditions. Already in early 1990s, we witnessed successful neural network applications; two that stand out were LeCun's *LeNet* network for recognizing handwritten digits and Tesauro's *TD-Gammon* network that played backgammon. In this last decade with more data and computing power available, we have been seeing many impressive applications of "deep" neural networks composed of sometimes hundreds of such layers. So, for example, networks deeper than *LeNet* are now being used in face recognition, and a network deeper than *TD-Gammon* has learned to play Go. The recent tectonic interest in machine learning and artificial intelligence is largely due to such successes with deep neural networks. Such successes unfortunately have also led to unrealistic claims and far-fetched extrapolations.

Machine learning is at the intersection of statistics and computer science, occasionally also taking inspiration from cognitive science and neuroscience. Research in these different communities followed different paths in the past with different emphases. Our aim in this book is to bring them together to give a unified introductory treatment of the field, together with some interesting applications, devoid, as much as possible, of hype.

Now, let us start discussing the basic concepts and applications of machine learning.

MACHINE LEARNING, STATISTICS, AND DATA ANALYTICS

Learning to Estimate the Price of a Used Car

We saw in the previous chapter that we use machine learning when we believe there is a relationship between the observations of interest but do not know how. Because we do not know its exact form, we cannot just go ahead and write down the computer program. So our approach is to collect data of example observations and to analyze it to discover the relationship. Now, let us discuss further what we mean by a relationship and how we extract it from data; as always, it is a good idea to use an example to make the discussion concrete.

Consider the problem of estimating the price of a used car. This is a good example of a machine learning application because we do not know the exact formula for this; at the same time, we know that there should be some rules:

the price depends on the properties of the car, such as its brand; it depends on usage, such as mileage; and it even depends on things that are not directly related to the car, such as the current state of the economy.

Though we can identify these as the factors, we do not know exactly how they affect the price. For example, we know that, on average, as mileage increases price decreases, but we do not know how quickly this occurs. How these factors are combined to determine the price is what we do not know; luckily, we have data to help us. We can look at a number of cars currently in the market, record their attributes and how much they go for, and then we can try to learn the specifics of the relationship between such attributes and the price.

In doing that, the first question is to decide what to use as the *input representation*, that is, the attributes that we believe have an effect on the price of a used car. Those that immediately come to mind are the make and model of the car, its year of production, and its mileage. You can think of others too, but such attributes should be easy to record.

One important fact is that there can be two different cars having exactly the same values for these attributes, yet they can still go for different prices. This is because there are other factors that have an effect, such as accessories. There may also be factors that we cannot directly observe and hence cannot include in the input—for example, all the conditions under which the car has been driven in the past and how well the car has been maintained.

The crucial point is that no matter how many properties we list as input, there are always other factors that affect the output; we cannot possibly record and take all of them as input, and all these other factors that we neglect introduce uncertainty.

The effect of this uncertainty is that we can no longer estimate an exact price, but we can estimate an *interval* in which the unknown price is likely to lie, and the length of this interval depends on the amount of uncertainty—it defines how much the price can vary due to those factors that we do not, or cannot, take as input.

Randomness and Probability

In mathematics and engineering, we model uncertainty using *probability theory*. In a deterministic system, given the inputs, the output is always the same; in a random process, the output depends also on uncontrollable factors that introduce randomness.

Consider the case of tossing a coin. It can be claimed that if we have access to knowledge such as the exact composition of the coin, its initial position, the amount, position, and the direction of the force applied to the coin when tossing it, where and how it is caught, and so forth, the outcome of the toss can be predicted exactly; but because all this information is hidden from us, we can only talk about the probability of the outcomes of the toss. We do not know if the outcome is heads or tails, but we can say something about the probability of each outcome, which is a measure of our belief in how likely that outcome is. For example, if a coin is fair, the probability of heads and the probability of tails are equal—if we toss it many times, we expect to see roughly as many heads as tails.

If we do not know those probabilities and want to *estimate* them, then we are in the realm of *statistics*. We follow the common terminology and call each data instance an "example" and reserve the word "sample" for the *collection* of such examples. The aim is to build a *model* to explain the process that generates the sample. In the coin tossing case, we collect a sample by tossing the coin a number of times and record the outcomes—heads or tails—as our observations. Then, our estimator for the probability of heads can simply be the proportion of heads in the sample—if we toss the coin six times and see four heads and two tails in our sample, we estimate the probability of heads as ⅔ (and hence the probability of tails as ⅓). Then if we are asked to guess the outcome of the next toss, our estimate will be heads because it is the more probable outcome.

This type of uncertainty underlies games of chance, which makes gambling a thrill for some people. But most of us do not like uncertainty, and we try to avoid it in our lives, at some cost if necessary. For example, if the stakes are high, we buy insurance—we prefer the certainty of never losing a large amount of money (due to accidental loss of something of high worth) to the cost of paying a small premium, even if the event is very unlikely.

The price of a used car is similar in that there are uncontrollable factors that make the depreciation of a car a random process. Two cars following one another on the production line are exactly the same at that point and hence are worth exactly the same. Once they are sold and start being used, all sorts of factors start affecting them: one of the owners may be more meticulous, one of the cars may be driven in better weather conditions, one car may have been in an accident, and so on. Each of these factors is like a random coin toss that varies the price.

A similar argument can also be made for customer behavior in retail. We expect customers in general to follow certain patterns in their purchases depending on factors such as the composition of their household, their tastes, their income, and so on. Still, there are always additional random factors that introduce variance: vacation, change in weather, some catchy advertisement, and so on. As a result of this randomness, we cannot estimate exactly which items will be purchased next, but we can calculate the probability that an item will be purchased. Then if we want to make predictions, we can just choose the items whose probabilities are the highest.

Learning a General Model

Whenever we collect data, we need to collect it in such a way as to learn general trends. For example, in representing a car, if we use the brand as an input attribute, we define a very specific car. But if we instead use general attributes such as the number of seats, engine power, trunk volume, and so forth, we can learn a more general estimator. This is because different models and makes of cars all appeal to the same type of customer, called a customer segment, and we would expect cars in the same segment to depreciate similarly. Ignoring the brand and concentrating on the basic attributes that define the segment is equivalent to using the same, albeit noisy, data instance for all such cars of the same type; it effectively increases the size of our data.

A similar argument can also be made for the output. Instead of estimating the price as is, it makes more sense to estimate the percentage of its original price, that is, the effect of depreciation. This again allows us to learn a more general model.

Though of course it is good to learn models that are general, we should not try to learn models that are too general. For example, cars and trucks have very different characteristics, and it is better to collect data separately and learn different models for the two than to collect data and try to learn a single model covering both.

Another important fact is that the underlying task may change in time. For example, the price of a car depends not only on the attributes of the car, the attributes representing its past usage, or the attributes of the owner, but also on the state of the economy, that is, the price of other things. If the economy, which is the environment in which we do all our buying and selling, undergoes significant changes, previous trends no longer apply. Statistically speaking, the properties of the random process that underlie the data have changed—we are given a new set of coins to toss. In this case, the previously learned model does not hold anymore, and we need to collect new data and learn again; or, if we have a mechanism for getting feedback about our performance, we can fine-tune the model as we continue to use it.

Model Selection

One of the most critical points in learning is the *model* that defines the template of the relationship between the inputs and the output. For example, if we believe that we can write the output as a weighted sum of the attributes, we can use a *linear model* where attributes have an additive effect.

Let us see an example. Assume we only take mileage as our input attribute and that we collect the data given in

Table 1 The example data set is composed of seven cars, each with its mileage (in miles) and price (US dollars). The index is just to name them; the order of the cars is not important.

Index	Mileage	Price
1	9,000	38,500
2	95,000	6,000
3	20,000	32,000
4	60,000	15,000
5	15,000	35,000
6	30,000	27,000
7	90,000	12,000

table 1. We have seven cars whose mileage and price values are recorded. We can fit a linear model to that data; that is, we draw the line that passes as close as possible to those data points. Figure 1 shows the data points and the fitted line. That linear model is written as

$$y = 39258 - 0.3427x$$

where x denotes the mileage and y denotes the estimated price. The line starts at 39,258 US dollars—that is the estimate for a car with 0 miles—and every additional mile decreases the price by 34.27 cents, or equivalently, every

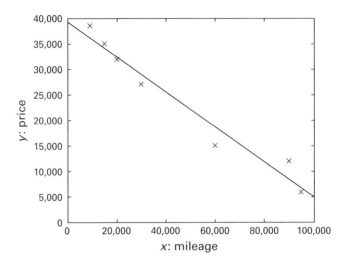

Figure 1 Estimating the price of a used car as a regression task. Each cross represents one car from table 1 where the horizontal *x*-axis is its mileage and the vertical *y*-axis is its price. Together they constitute the training set. In estimating the price of a used car, we want to learn a model that fits (passes as close as possible to) these data points; an example linear fit is shown. Once such a model is fit, it can be used to predict the price of any car given its mileage.

additional 10,000 miles driven pulls the price down by 3,427 US dollars.

That value, −0.3427, is the *weight* of mileage on price. Not only the value but the sign of a weight is also informative. Here, analysis of the data indicated that the relationship is negative—as mileage increases price decreases—but in another example the weight can be positive; for example, if we have an additional attribute for engine power, we will find that price increases with larger engines. It may also be the case that the weight estimated from the data turn out to be very close to zero; in such a case, we can conclude that the corresponding attribute is not important and eliminate it from the model.

These weights are the *parameters* of the model and are calculated from the data. The model is always fixed; it is the parameters that are adjustable, and it is this process of adjustment to better match the data that we call learning.

The linear model is very popular because it is simple; it has few parameters and it is easy to calculate a weighted sum. It is easy to understand and interpret. Furthermore, it works surprisingly well for a lot of different tasks.

No matter how we vary its parameters, each model can only be used to learn a set of problems and *model selection* refers to the task of choosing between models. Selecting the right model is a more difficult task than optimizing the parameters of a given model, and information about the application is helpful.

For instance, here, in estimating car prices, the linear model may not be applicable if the range is long. It has been seen empirically that the effect of the age is not arithmetic but geometric: each additional year does not decrease the price by the same amount, but a typical vehicle loses 15 percent of its value each year.[1] In later sections, we discuss some machine learning algorithms that use nonlinear models that are more powerful, in the sense that they can be used in a larger variety of applications.

Here, it may be a good idea to point out the difference between machine learning and the usual treatment of data in databases. Table 1 may have been drawn from a database, containing probably not only the mileage and the price, but all sorts of other information about the cars and maybe also their owners. Given such a database, one can make a query such as, "What is the price of the car with 20,000 miles?" and find that it is 32,000 US dollars. With a database, we only have information and care about those particular cars; a query such as "What is the price of the car with 25,000 miles?" is not meaningful because there is no such car in the database.

In the case of machine learning, or statistics in general, we consider the data in table 1 as a *sample*, drawn from the *population* of all used cars. We assume that there is an underlying process whereby cars lose value as they are driven, but we do not know how this process works, and we want to extract it from examples. We cannot possibly access all possible cars in the population and record the mileage and price of all of them; the ones we have access to make up a small subset which is our sample. From a statistical point of view, we do not particularly care about this particular sample, there can be mistakes in it, or it could just have been a different sample; what we care about is the population from which any sample is drawn. And the advantage of learning, namely, going from the particular sample to the general population, is that now we can query it with 25,000 miles, which is equivalent to asking, "What would be the price of a *typical* car with 25,000 miles?" where "typical" means averaged over all used cars with 25,000 miles.

Supervised Learning

This task of estimating a numeric output value from a set of input values is called *regression* in statistics; for the linear model, we have linear regression. In machine learning, regression is one type of *supervised learning*. In this type of learning, for each example we have the input and the desired output. The name comes from the supposition that there is a *supervisor* who can provide us with the desired output for any input. When we collect data by looking at the cars currently sold in the market, we are able to observe both the attributes of the cars and also their prices.

Earlier, we used the linear model with its weight parameters. Each model corresponds to a certain type of dependency assumption between the inputs and the output. Learning corresponds to adjusting the parameters so that the model makes the most accurate predictions on the data. In the general case, learning implies getting better according to a performance criterion, and in regression, performance depends on how close the model predictions are to the observed output values in the training data. The assumption here is that the training data is large and diverse enough to cover sufficiently well the characteristics of the underlying task, so a model that works accurately on the training data can be said to have learned the task.

The different machine learning algorithms we have in the literature either differ in the models they use, or in the performance criteria they optimize, or in the way the parameters are adjusted during this optimization.

At this point, we should remember that the aim of machine learning is rarely to replicate the training data but the correct prediction of new cases. If there were only a certain number of possible cars in the market and if we knew the price for all of them, then we could simply store all those values and do a table lookup; this would be *memorization*. But frequently (and this is what makes learning interesting), we see only a small subset of all possible instances and from that data, we want to *generalize*—that is, we want to learn a general model that goes beyond the training examples to also make good predictions for inputs not seen during training.

Having seen only a small subset of all possible cars, we would like to be able to predict the right price for a car outside the training set, one for which the correct output was not given in the training set. How well a model trained on the training set predicts the right output for such new instances is called the *generalization ability* of the model and the learning algorithm.

The basic assumption we make here (and it is this assumption that makes learning possible) is that similar cars have similar prices, where similarity is measured in terms of the input attributes we choose to use. As the values of these attributes change slowly—for example, as mileage changes—price is also expected to change slowly. There is smoothness in the output in terms of the input, and that is what makes generalization possible. Without such regularity, we cannot go from particular cases to a general model, as then there would be no basis in the belief that there can be a general model that is applicable to all cases, both inside and outside the training set.

Not only for the task of estimating the price of a used car, but for many tasks where data is collected from the real world, be they for business applications, pattern recognition, or science, we see this smoothness. Machine learning, and prediction, is possible because the world has regularities. Things in the world change smoothly. This is Leibniz's dictum that "Nature does not make jumps." We are not "beamed" from point A to point B, but we need to pass through a sequence of intermediate locations. Objects occupy a continuous block of space in the world. Nearby points in our visual field belong to the same object

Machine learning, and prediction, is possible because the world has regularities. Things in the world change smoothly. This is Leibniz's dictum that "Nature does not make jumps.

Figure 2 The Kanizsa triangle, originally created by the Italian psychologist Gaetano Kanizsa in 1955. Although we do not see its whole contour but just its tips, we imagine a white triangle in the foreground partially occluding the three black circles, because such a triangle is the simplest explanation for what we see.

and hence mostly have shades of the same color. Sound too, whether in song or speech, changes smoothly. Discontinuities correspond to boundaries, and they are rare. Most of our sensory systems make use of this smoothness; what we call visual illusions, such as the Kanizsa triangle (see figure 2), are due to the smoothness assumptions of our sensory organs and brain.[2]

Such an assumption is necessary because collected data is not enough to find a unique model—learning, or fitting a model to data, is an *ill-posed problem*. Every learning algorithm makes a set of assumptions about the data to find a unique model, and this set of assumptions is called the *inductive bias* of the learning algorithm (Mitchell 1997).

This ability of generalization is the basic power of machine learning; it allows going beyond the training instances. Of course, there is no guarantee that a machine learning model generalizes correctly—it depends on how suitable the model is for the task, how much training data there is, and how well the model parameters are optimized—but if it does generalize well, we have a model that is much more than the data. This is how we assess learning: a student who can solve only the exercises that the teacher previously solved in class has not fully mastered the subject; we want them to acquire a sufficiently general understanding from those examples so that they can also solve new questions about the same topic.

Learning a Sequence

Let us see a very simple example. You are given a sequence of numbers and asked to find the next number in the sequence. Let us say the sequence is

0, 1, 1, 2, 3, 5, 8, 13, 21, 34, 55

You probably noticed that this is the *Fibonacci sequence*. The first two terms are 0 and 1, and every term that follows is the sum of its two preceding terms, for example, $8 + 13 = 21$. Once you identify the model, you can use it to make a prediction and guess that the next number will be $34 + 55 = 89$, and then $55 + 89 = 144$, and so on. You can then keep predicting using the same model and generate a sequence as long as you like.

The reason we come up with this answer is that we are unconsciously trying to find a *simple explanation* for this data. This is what we always do. In philosophy, *Occam's razor* tells us to prefer simpler explanations, eliminating unnecessary complexity. For this sequence, a linear rule where two preceding terms are added is simple enough.

If the sequence were shorter,

0, 1, 1, 2

you would not immediately go for the Fibonacci sequence— my prediction would be 2. With short sequences, there are many possible, and simpler, rules. As we see one more piece of data, those rules whose next value does not match are eliminated. Model fitting is basically a process of elimination: each extra observation (training example) is a constraint that eliminates all those candidates that do not obey it. And once we run out of the simple ones, we need to move on to complicated explanations incrementally to cover all the terms.

The complexity of the model is defined using *hyperparameters*. Here the fact that the model is linear and that only the previous two terms are used are hyperparameters.

Let us say the sequence is

0, 1, 1, 2, 3, 7, 16, 65, 321

the rule is to sum the *square* of the first and the second; for example, $7 \times 7 + 16 = 65$. This is not linear but quadratic, which is a more complex rule—there is an additional multiplication.

If the sequence is

0, 1, 1, 2, 4, 7, 13, 24, 44

the rule is to sum the *three* preceding values, given the initial three values 0, 1, 1; for example, $4 + 7 + 13 = 24$. This is also more complex than the Fibonacci proper because it uses three values (inputs) as opposed to two. As you see, one needs to adjust the complexity of the model to that of the task underlying the data; we can do this either by changing the complexity of the calculation or by taking more inputs into account.

Now let us say that the sequence is

0, 1, 1, 2, 3, 6, 8, 13, 20, 34, 55

Maybe you can also find a rule that explains this sequence, but I imagine it will be a complicated one. The alternative is to say that this is the Fibonacci sequence *with two errors* (6 instead of 5 and 20 instead of 21) and still predict 89 as the next number—or we can predict that the next number will lie in the interval [88,90].

Instead of a complex model that explains this sequence exactly, a noisy Fibonacci may be a more likely explanation if we believe that there may be errors (remember our earlier discussion on random effects due to unknown factors). And indeed, errors are likely. Most of our sensors are far from perfect, typists make typos all the time, and though we like to believe that we act reasonably and rationally, we also act on a whim and buy/read/listen/click/travel on impulse.

Learning also performs *compression*. Once you learn the rule underlying the sequence, you do not need the data anymore. By fitting a rule to the data, we get an explanation that is simpler than the data, requiring less memory to store and less computation to process. Once we learn the rules of multiplication, we do not need to remember the product of every possible pair of numbers.

Credit Scoring

Let us now see another application to help us discuss another type of machine learning algorithm. A *credit* is an

By fitting a rule to the data, we get an explanation that is simpler. Once we learn the rules of multiplication, we do not need to remember the product of every possible pair of numbers.

amount of money loaned by a financial institution such as a bank, to be paid back with interest, generally in installments. It is important for the bank to be able to predict in advance the risk associated with a loan, which is a measure of how likely it is that the customer will default and not pay the whole amount back. This is both to make sure that the bank will make a profit and also to not inconvenience a customer with a loan over their financial capacity.

In *credit scoring*, the bank calculates a risk given the amount of credit and the information about the customer. This information includes data we have access to and is relevant in calculating the customer's financial capacity—namely, income, savings, collateral, profession, age, past financial history, and so forth. Again, there is no known rule for calculating the score; it changes from place to place and time to time. So the best approach is to collect data and try to learn it.

Credit scoring can be defined as a regression problem; historically the linear model, where the score of a customer was written as a weighted sum of different attributes, was frequently used. Each additional thousand dollars in salary increases the score by S points, and each additional thousand dollars of debt decreases the score by D points, where again the parameters S and D can be learned from data. Once we have such a model, we can use it on a new application to make a decision where depending on the estimated score, different actions can be taken—for example, a customer with a higher score may have a higher limit on their credit card.

Instead of regression, credit scoring can also be defined as a *classification* problem, where there are the two classes of customers: low-risk and high-risk. Classification is another type of supervised learning where the output is a class code, as opposed to the numeric value we have in regression.

A *class* is a set of instances that share a common property, and in defining this as a classification problem, we are assuming that all high-risk customers share some common characteristics not shared by low-risk customers, and that there exists a formulation of the class in terms of those characteristics, called a *discriminant*. We can visualize the discriminant as a boundary separating examples of the two classes in the space defined by the customer attributes.

As usual, we are interested in the case where we do not know the underlying discriminant but have a sample of example data, and we want to learn the discriminant from data.

In preparing the data, we look at our past records, and we label the customers who paid back their loans as low-risk and those who defaulted as high-risk. Analyzing this data, we would like to learn the class of high-risk customers so that in the future, when there is a new application for a loan, we can check whether or not the customer matches that description and reject or accept the application accordingly.

The information about a customer makes up the input to the classifier whose task is to assign the input to one of the two classes. Using our knowledge of the application, let us say that we decide to take a customer's income and savings as input (see figure 3). We observe them because we have reason to believe that they give us sufficient information about the credibility of a customer.

One possible model defines the discriminant in the form of if-then rules:

IF income < X AND savings < Y THEN high-risk ELSE low-risk

where X and Y are the parameters fine-tuned to the data, to best match the rule prediction with what the data tells us (see figure 3).

In this model the parameters are these thresholds, not weights as we have in the linear model. In regression, the task is to find a line that passes as close as possible to the data points; in classification, it is to fit a separating boundary between the data points from different classes.

Each if-then rule specifies one composite condition made up of terms, each of which is a simple condition on one of the input attributes. The antecedent of the rule

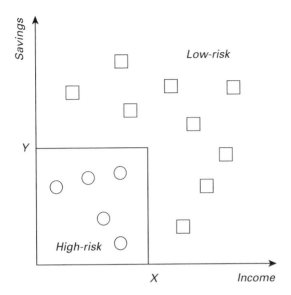

Figure 3   Separating the low- and high-risk customers as a classification problem. The two axes are the income and savings, each in its unit (e.g., in thousands of dollars). Each customer, depending on their income and savings, is represented by one point in this two-dimensional space, and their class is represented by shape—a circle represents a high-risk customer and a square represents a low-risk customer. All the high-risk customers have their income less than X and savings less than Y, and hence this condition can be used as a discriminant, whose shape is shown using thick lines.

is an expression containing terms connected with AND, namely, a conjunction; that is, all the conditions should hold for the rule to apply.

We understand from the rule that among the subset of customers that satisfies the antecedent—namely, those whose income is less than X *and* whose savings is less than Y—there are more high-risk than low-risk customers, so the probability of high risk for them is higher; that is why the consequent of the rule has high-risk as its label.

In this simple example, there is a single way of being high-risk and all the remaining cases are low-risk, so one rule is sufficient. In another application, there may be a *rule base* that is composed of several if-then rules, each of which delimits a certain region, and each class is specified using a disjunction of such rules. There are different ways of being high-risk, each of which is specified by one rule and satisfaction of any of the rules is enough.

Learning such rules from data allows *knowledge extraction*. The rule is a simple model that explains the data and looking at this model we have an explanation about the process underlying the data. For example, once we learn the discriminant separating the low-risk and high-risk customers, we have knowledge about the properties of low-risk customers. We can then use this information to target potential low-risk customers more efficiently, such as through customized advertising.

Expert Systems

Before machine learning was the norm, expert systems existed. Proposed in 1970s and used in 1980s,[3] they were computer programs that aided humans in decision making.

An *expert system* is composed of a *knowledge base* and an *inference engine*. The knowledge is represented as a set of if-then rules, like the ones we discussed earlier, and the inference engine uses logical inference rules for deduction. The rules are programmed after consultation with domain experts, and they are fixed. This process of converting domain knowledge to if-then rules was manual, and hence difficult and costly. The inference engine was programmed in specialized programming languages such as Lisp and Prolog, which are especially suited for logical inference.

For a time in the 1980s, expert systems were quite popular around the world, not only in the United States (where Lisp was used), but also in Europe (where Prolog was used). Japan had a Fifth Generation Computer Systems Project for massively parallel architectures for expert systems and artificial intelligence (AI). There were applications, but in rather limited domains, such as MYCIN for diagnosing infectious diseases (Buchanan and Shortliffe 1984); commercial systems also existed.

Despite the research and the wide interest, expert systems never took off. There are basically two reasons for this. First, the knowledge base needed to be created manually through a very laborious process; there was no learning from data. The second reason was the unsuitability of logic to represent the real world. In real life, things are not true or false, but have grades of truth: a person is not either old or not old, but oldness increases gradually with age. The logical rules too may apply with different degrees of certainty: "If X is a bird, X can fly" is mostly true but not always.

To represent degrees of truth, *fuzzy logic* was proposed with fuzzy memberships, fuzzy rules, and inference, and since its inception, had some success in a variety of applications (Zadeh 1965). Another way to represent uncertainty is to use probability theory, as we do in this book.

Machine learning systems that we discuss in this book are extensions of expert systems in decision making in two ways: first, they need not be programmed but can learn from examples, and second, because they use probability theory, they are better in representing real-world settings with all the concomitant noise, exceptions, ambiguities, and resulting uncertainty.

Expected Values

When we make a decision—for example, when we choose one of the classes—we may be correct or wrong. It may be the case that correct decisions are not equally good and wrong decisions are not equally bad. When making a decision about a loan applicant, a financial institution should take into account both the potential gain as well as the potential loss. An accepted low-risk applicant increases profit, while a rejected high-risk applicant decreases loss. A high-risk applicant who is erroneously accepted causes loss, and an erroneously rejected low-risk applicant is a missed chance for profit.

The situation is much more critical and far from symmetrical in other domains, such as *medical diagnosis*. Here, the inputs are the relevant information we have about the patient and the classes are the illnesses. The inputs contain the patient's age, gender, past medical history and current symptoms. Some tests may not have been applied to the patient, and these inputs would be missing. Tests take time, are costly, and may inconvenience the patient, so we do not want to apply them unless we believe they will give us valuable information.

In the case of medical diagnosis, a wrong decision may lead to wrong or no treatment, and the different types of error are not equally bad. Let us say we have a system that collects information about a patient and based on those, we want to decide whether the patient has a certain disease (say a certain type of cancer) or not. There are two possibilities: either the patient has cancer—let us call it the positive class—or not—the negative class.

Similarly, there are two types of errors (see table 2). If the system predicts cancer but in fact the patient does not have it, this is a *false positive*—the system chooses the positive class wrongly. This is bad because it will cause unnecessary treatment, which is both costly and inconvenient for the patient. If the system predicts no disease when in fact the patient has it, this is a *false negative*. A false negative has a higher cost than a false positive because then the patient will not get the necessary treatment. Because the cost of a false negative is so much larger than the cost of a false positive, we would choose the positive class—to start a closer investigation—even if the probability of the positive class is relatively small. This is not like predicting a coin toss where we choose the outcome—heads or tails—whose probability is higher than ½ (because the two possible wrong guesses are equally bad).

Table 2 Different types of errors in decision making

	Action		
Truth	*Choose positive (start treatment)*	*Choose negative (discharge the patient)*	*Sum*
Positive (the patient has cancer)	TP: True positive	FN: False negative	P
Negative (the patient does not have cancer)	FP: False positive	TN: True negative	N
Sum	P′	N′	

This is the basis of *expected value* calculation where not only do we decide by using probabilities, but we also take into account the possible loss or gain we may face as a result of our decision. Though expected value calculation is frequently done in many domains, such as in insurance, it is known that people do not always act rationally; if that were the case, no one would buy a lottery ticket!

In Max Frisch's novel *Homo Faber*, the mother of a girl who was bitten by a snake is told not to worry because the mortality from snakebites is only 3–10 percent. The woman gets angry and says, "If I had a hundred daughters . . . then I should lose only three to ten daughters. Amazingly few! You're quite right," and then she continues, "I've only got one child." We need to be careful in using expected value calculation when ethical matters are involved; later on for the case of self-driving cars, we will get back to ethical and also legal aspects of automated decision making.[4]

If both false positive and false negative have high costs, a possible third action is to *reject* and defer decision. For example, if computer-based diagnostics cannot choose between two outcomes, it can opt to reject, and the case can be decided manually; the human expert can make use of additional information not directly available to the system. In an automated mail sorter, if the system cannot recognize the numeric zip code on an envelope, the postal worker can also read the address.

In classification, when false positives and false negatives are equally bad, their sum, FP + FN, gives us the number of misclassifications. For example, let us say we have a visual recognition system that separates cars from tanks, and let us assume cars make up the positive class. Then FP is the number of tanks classified as cars and FN is the number of cars classified as tanks. Their sum divided by the total number of images, $(FP + FN)/(P + N)$, is the *classification error*, or equivalently, $(TP + TN)/(P + N)$, is the *classification accuracy*.

In other applications, the performance criterion can be different. Let us envisage an application where people are allowed to access their bank accounts by their voice. A false positive is an allowed impostor and a false negative is a valid user that is denied service. The former is much worse than the second. The *hit rate*, or the *true-positive rate*, measures what proportion of valid users is correctly authenticated, TP/P, and *false alarm rate*, or *false-positive rate*, measures what proportion of impostors are wrongly authenticated, FP/N.

There is a trade-off between the two. If we modify our classifier so that it chooses the positive class more easily, this increases the hit rate, but also risks increasing the false alarm rate. In deciding when to choose which class, we want to make the hit rate as large as possible while keeping the false alarm rate as small as possible.

In *information retrieval*, we have a query—for example, defined using keywords—and we want to retrieve records that match the query from a database. For example, let us say we want to retrieve images of tigers using the keyword "tiger" from a database of images. In such a case, an image that is retrieved corresponds to assignment to the positive class and an image that is not retrieved corresponds to assignment to the negative class. In this case, TP corresponds to the number of tiger images that are retrieved, and FN to the number of tiger images that exist in the database but are not retrieved. FP is the number of images that are retrieved but are not tiger images.

In this scenario, we have two performance criteria. *Precision* is the ratio of true positives to all the retrieved instances, namely, $TP/P' = TP/(TP + FP)$, that is, what percentage of the retrieved instances are really relevant, that is, match the query. *Recall* is the ratio of true positives to all the positive instances, namely, TP/P, that is, what percentage of the relevant instances are actually retrieved.

We want both *precision and recall* to be as close to 1 as possible: If precision is 1, all the retrieved records may be relevant but there may still be relevant records that are not retrieved. If recall is 1, all the relevant records are retrieved but there may also be other retrieved irrelevant records (see figure 4). Again, we see the trade-off between

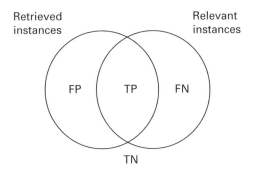

Figure 4   Precision and recall explained in terms of Venn diagrams.
Precision is TP/(TP + FP) and recall is TP/(TP + FN).

precision and recall. We can always increase recall by retrieving more images but that risks decreasing precision.

We see that unlike in classification, we do not care about the true negatives here; those are non-tiger images that are correctly not retrieved; in retrieval we do not care about those, their number can increase or decrease depending on the size and scope of the database without affecting our performance assessment.

In other domains—for example, in medicine—people use the measures of *sensitivity* and *specificity*. *Sensitivity* is the same as recall, which measures how well we detect the positives. *Specificity* is how well we detect the negatives, which is TN/N—it is equal to 1 – false alarm rate. For example, let us say we have developed a test for a virus: Sensitivity is how well the test catches the people who have the virus and specificity is how well it is silent on people who do not have the virus.

In this chapter, we have discussed the basics of machine learning in general; in the next chapter, we will cover one type of learning, supervised learning, that finds use, for example, in recognizing patterns such as faces and speech.

PATTERN RECOGNITION

Learning to Read

Different automatic visual recognition tasks have different complexities. One of the simplest is the reading of *barcodes* where information is represented in terms of lines of different widths, which are shapes that are easy to recognize. The barcode is a simple and efficient technology: It is easy to print barcodes, and it is also easy to build scanners to read them; that is why they are still widely used. But the barcode is not a natural representation, and the information capacity is limited; recently two-dimensional matrix barcodes have been proposed where more information can be coded in a smaller area; for example, QR codes that can be scanned by a smartphone point to a website.

There is always a trade-off in engineering. When the task is difficult to solve, we can devise more efficient solutions by constraining it. For example, the wheel is a very good solution for transportation, but it requires flat surfaces and so roads too have to be built. The controlled environment makes the task easier. Legs work in a variety of terrains, but they are more difficult to build and control, and they can carry a significantly less heavy load.

Optical character recognition is recognizing printed or written characters from their images. This is more natural than barcodes because no extra coding (in terms of bars) is used. If a single font is used, there is a single way of writing each character; there are standardized fonts such as OCR-A, defined specifically to make automatic recognition easier.

With barcodes or a single font, a single template exists for each class and there is no need for learning. For each character, we have a single prototype that we can simply store. It is the ideal image for that character, and we compare the seen input with all the prototypes one by one and choose the class with the best matching prototype—this is called *template matching*. There may be errors in printing or sensing, but we can do recognition by finding the closest match.

If we have many fonts or handwritings, we have multiple ways of writing the same character, and we cannot possibly store all of them as possible templates. Instead, we want to "learn" the class by going over all the different examples of the same character and find some general description that covers all of them.

It is interesting that though writing is a human invention, we do not have a formal description of 'A' that covers all 'A's and none of the non-'A's. Not having it, we take samples from different writers and fonts, and learn a definition of 'A'-ness from these examples. But though we do not know what it is that makes an image an instance of the class 'A', we are certain that all those distinct 'A's have something in common, which is what we want to extract from the examples.

We know that a character image is not just a collection of random dots and strokes of different orientations, but it has a regularity that we believe we can capture by using a learning program. For each character, we see examples in different fonts (for printed text) or writings (for handwritten text) and generalize; that is, we find a description that is shared by all of the examples of a character: 'A' is one way of combining a certain set of strokes, 'B' is another way, and so on.

Recognition of printed characters is relatively easy with the Latin alphabet and its variants; it is trickier with alphabets where there are more characters, accents, and writing styles. In cursive handwriting, characters are connected and there is the additional problem of segmentation.

Many different fonts exist, and people have different handwriting styles. Characters may also be small or large, slanted, printed in ink or written with a pencil, and as a result, many possible images can correspond to the same character. Despite all the research, there is still no computer program today that is as accurate as humans for this task. That is why captchas are still used, a *captcha* being a corrupted image of words or numbers that needs to be typed to prove that the user is a human and not a computer program.

Matching Model Granularity

In machine learning, the aim is to fit a model to the data. In the ideal case, we have one single, *global* model that applies to all of the instances. For all cars, as we saw in chapter 2, we have a single regression model that we can use to estimate the price. In such a case, the model is trained with the whole training data and all the instances have an effect on the model parameters. In statistics, this is called *parametric estimation*.

The parametric model is good because it is simple—we store and calculate a single model—and it is trained with the whole data. Unfortunately, it may be restrictive in the sense that this assumption of a single model applicable to all cases may not hold in all applications. In certain tasks, we may have a set of *local* models, each of which is applicable to a certain type of instances. This is *semi-parametric estimation*. We still have a model that maps the input to the output but is valid only locally, and for different type of inputs we have different models.

For example, in estimating the price of used cars, we may have one model for sedans, another model for sports cars, and another for luxury cars, if we have reason to believe that for these different types of cars, the depreciation behaviors are different. In such an approach, each local model is trained only with the training data that falls within its scope—the number of local models is the hyperparameter defining the model flexibility and hence complexity.

In certain applications, even the semi-parametric assumption may not hold; that is, the data may lack a clear structure and it cannot be explained in terms of a few local models. In such a case, we use the other extreme of *nonparametric estimation* where we assume no simple model, either globally or locally. The only information we use is the most basic assumption—namely, that similar inputs have similar outputs. In such a case, we do not have an explicit training process that converts training data to model parameters; instead, we just keep the training data as the sample of past cases.

Given an instance, we find the training instances that are most similar to the query and we calculate an output in terms of the known outputs of these past similar instances. For example, given a car whose price we want to estimate, we find among all the training instances the three cars that are most similar—in terms of the attributes we use—and then calculate the average of the prices of these three cars as our estimate. Because those are the cars that are most similar in their attributes, it makes sense that their prices should be similar too. This is called *k-nearest neighbor estimation* where here *k* is three. Since those are the three most similar past "cases," this approach is sometimes called *case-based reasoning*. The nearest-neighbor algorithm is intuitive: similar instances mean similar things. We all love our neighbors because they are so much like us—or we hate them, as the case may be, for exactly the same reason.

Generative Models

An approach that has recently become very popular in data analysis is to consider a *generative model* that represents our belief as to how the data is generated. We assume that there is a hidden model with a number of *hidden*, or *latent*, *causes* that interact to generate the data we observe. Though the data we observe may seem big and complicated, it is produced through a process that is controlled by a few variables, which are the hidden factors, and if we can somehow infer these, the data can be represented and understood in a much simpler way. Such a simple model, if it is appropriately chosen and well trained, can also make accurate predictions.

Consider optical character recognition. Generatively speaking, we can say that each character image is composed of two types of factors: there is the identity, namely the label of the character, and there is the appearance, those that are due to the process of writing or printing.

In a printed text, the appearance part may be due to the font; for example, characters in Times Roman font have serifs and strokes that are not all of the same width. The font is an aesthetic concern; in calligraphy, it is the aesthetic part that becomes especially prominent. Just like the choice of font in printed text, the handwriting style of the writer introduces variance in written text. But these added characteristics due to appearance should not be large enough to cause confusion about the identity, otherwise we say that the person has a bad or illegible handwriting. The appearance also depends on the material the writer is using (e.g., pen versus pencil) and also on the medium (e.g., paper versus marble slab).[1]

The printed or written character may be large or small, and this is generally handled at a preprocessing stage of normalization where the character image is converted to a fixed size—we know that the size does not affect the identity. This is called *invariance*. We want invariance to size (whether the text is 12pt or 18pt, the content is the same) or invariance to slant (as when the text is in italics) or invariance to the width of strokes (as in bold characters). But, for example, we do not want invariance to large rotations: *q* is a rotated *b*.

In recognizing the class, we need to focus on the identity, and we should find attributes that represent the identity, and learn how to combine them to represent the character. We treat all the attributes that relate to the appearance, namely the writer, aesthetics, medium, and sensing, as irrelevant and learn to ignore them. But note that in a different task, those may be the important ones; for example, in authentication of handwriting or in signature recognition, it is the writer-specific attributes that become important and not the content.

The generative model is *causal* in that it explains how the data is generated by hidden factors that cause it. Once we have such a model trained, we may want to use it for *diagnostics*, which implies going in the opposite direction, that is, from observation to cause. Medical domain is a good example here: the diseases are the causes and they are hidden; the symptoms are the observed attributes of the patient, such as the results of medical tests. Going from disease to symptom is the causal direction—that is what the disease does; going from symptom to disease is diagnostics—that is what the doctor does. In the general

Going from disease to symptom is the causal direction—that is what the disease does; going from symptom to disease is diagnostics—that is what the doctor does.

case, diagnostics is the inference of hidden factors from observed variables.

A generative model can be represented as a graph composed of nodes that correspond to hidden and observed variables, and the arcs between nodes represent dependencies between them, such as causalities. Such *graphical models* are interesting in that they allow a visual representation of the problem, and statistical inference and estimation procedures can be mapped to well-known graph operations for which we already have efficient procedures in computer science (Koller and Friedman 2009).

In a graphical model, a causal link goes from a hidden factor to an observed symptom, while a diagnostics effectively inverts the direction of the link. We use conditional probability to model the dependency, and for example, when we talk about the conditional probability that a patient has a runny nose given that they have the flu, we go in the causal direction: the flu causes the runny nose (with a certain probability).

If we have a patient and we know they have a runny nose, we need to calculate the conditional probability in the other direction—namely, the probability that they have the flu given that they have a runny nose (see figure 5). In probability, the two conditional probabilities are related because of the *Bayes' rule*,[2] and that is why graphical models are sometimes also called *Bayesian networks*. In a later section, we return to Bayesian estimation; we

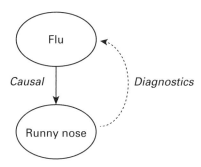

Figure 5   The graphical model showing that the flu is the cause of a runny nose. If we know that the patient has a runny nose and want to check the probability that they have the flu, we are doing diagnostics by making inference in the opposite direction (using Bayes' rule). We can form larger graphs by adding more nodes and links to show increasingly complicated dependencies.

see that we can also include the model parameters in such networks and that this allows additional flexibility.

If we are reading a text, one factor we can make use of is the language information. A word is a sequence of characters, and we rarely write an arbitrary sequence of characters; we choose a word from the lexicon of the language. This has the advantage that even if we cannot recognize a character, we can still read t?e word. Such contextual dependencies may also occur at higher levels, between words and sentences as defined by the syntactic and semantic rules of the language. Machine learning algorithms help us learn such dependencies for natural language processing, as we discuss shortly.

Face Recognition

In the case of *face recognition*, the input is the image captured by a camera and the classes are the people to be recognized. The learning program should learn to match the face images to their identities. This problem is more difficult than optical character recognition because the input image is larger, a face is almost three-dimensional, and differences in pose and lighting cause significant changes in the image. Certain parts of the face may also be occluded; glasses may hide the eyes and eyebrows, and a beard may hide the chin.

Just as in character recognition, we can think of two sets of factors that affect the face image: there are the features that define the identity, and there are features that have no effect on the identity but affect appearance, such as hairstyle; or expression (namely, neutral, happy, angry, and so forth). These appearance features may also be due to hidden factors that affect the captured face image, such as the source of illumination or the pose. If we are interested in the identity, we want to learn a face description that uses only the first type of features, learning to be invariant to features of the second type.

However, we may be interested in the second type of features for other tasks. Recognizing facial expressions allows us to recognize emotions, as opposed to identity. For example, during a video monitoring a meeting, we may want to keep track of the mood of the participants. Likewise, in online education, it is important to understand whether the student is confused or gets frustrated, to better adjust the speed of presenting the material. In *affective computing*, which is a field that is rapidly becoming popular, the aim is to have computer systems that can recognize and take into account human affects, that is, the observed manifestations of emotions.

If the aim is identification or authentication of people—for example, for security purposes—using the face image is only one of the possibilities. *Biometrics* is recognition or authentication of people using their physiological and/or behavioral characteristics. In addition to the face, examples of physiological characteristics are the fingerprint, iris, and palm; examples of behavioral characteristics include the dynamics of signature, voice, gait, and keystroke. For more accurate decisions, inputs from different modalities can be integrated. When there are many different input sources—as opposed to the usual identification procedures of photo, printed signature, or password—forgeries (spoofing) becomes more difficult and the system more accurate, hopefully without too much inconvenience to the users.

Speech Recognition

In *speech recognition*, the input is the acoustic signal captured by a microphone and the classes are the words that can be uttered. This time the association to be learned is between an acoustic signal and a word of some language.

Just as we can consider each character image to be composed of basic primitives like strokes of different orientations, a word is made up of *phonemes*, which are the basic speech sounds. In the case of speech, the input is *temporal*; words are uttered in time as a *sequence* of these phonemes, and some words are longer than others.

Different people, because of differences in age, gender, or accent, pronounce the same word differently, and again, we may consider each word sound to be composed of two sets of factors, those that relate to the word and those that relate to the speaker. Speech recognition uses the first type of features, whereas speaker authentication uses the second. Incidentally, this second type of features (those relating to the speaker) is not easy to recognize or to artificially generate—that is why the output of speech synthesizers still sounds "robotic."[3]

Just as in biometrics, researchers here also rely on the idea of combining multiple sources. For example, to recognize speech, in addition to the acoustic information, we can also use the video image of the speaker's lips and the shape of the mouth as they speak the words.

Natural Language Processing and Translation

In speech recognition, as in optical character recognition, the integration of a *language model* taking contextual information into account helps significantly. Decades of research on programmed rules in computational linguistics have revealed that the best way to come up with a language model (defining the lexical, syntactic, and semantic rules of the language) is by learning it from some large corpus of example data. The applications of machine learning to *natural language processing* are constantly increasing; see Hirschberg and Manning 2015 for a recent survey, or Eisenstein 2019 for a textbook on the topic.

One of the easier applications is *spam filtering*, where spam generators on one side and filters on the other side keep finding more and more ingenious ways to outdo each other. This is a classification problem with two classes, spam and legitimate emails. A similar application is *document categorization* where we want to assign text documents to one of several categories, such as arts, culture, politics, and so on.

A face is an image and a spoken sentence is an acoustic signal, but what is in a text? A text is a sequence of characters, but characters are defined by an alphabet and the relationship between a language and the alphabet is not straightforward. The human language is a very complex form of information representation with its lexical, syntactic, and semantic rules at different levels, together with its subtleties such as humor and sarcasm, not to mention the fact that a sentence almost never stands or should be interpreted alone, but is part of some dialogue or general context.

The easiest method for representing a text is the *bag of words* representation where we predefine a large vocabulary of words and then we represent each document by using a list of the words that appear anywhere in the document. That is, of the words we have chosen, we note which ones appear in the document and which ones do not. We lose the position of the words in the text, which may be good or bad depending on the application. In choosing a vocabulary, we choose words that are indicative of the task; for example, in spam filtering, words such as "opportunity" and "offer" are discriminatory. There is a preprocessing step where suffixes (e.g., "-ing," "-ed") are removed, and where noninformative words (e.g., "the," "of") are ignored.

Recently, analyzing messages on social media has become an important application area of machine learning. Analyzing posts to extract *trending topics* is one: this implies a certain novel combination of words that has suddenly started to appear a lot. Another task is *sentiment analysis*, that is, determining whether a customer is happy or not with a product (e.g., a politician). For this, one can define a vocabulary containing words indicative of the two classes—happy versus not happy—using the bag of words representation and learn how they affect the class descriptions.

Perhaps the most impressive application of machine learning is *machine translation*. After decades of research on hand-coded translation rules, it has become apparent that the most promising approach is to provide a very large sample of pairs of texts in both languages and to have a learning program automatically figure out the rules to map one to the other. In bilingual countries such as Canada, and in the European Union with its many official languages, it is relatively easy to find the same text carefully translated in two or more languages. Such data is used by machine learning approaches to translation.

In chapter 4, we will discuss *deep learning*, which shows a lot of promise for this task, in automatically learning the different layers of abstraction that are necessary for processing natural language.

Combining Multiple Models

In any application, we can use any one of various learning algorithms and instead of trying to choose the single best one, a better approach may be to use them all and combine their predictions. We saw before that each algorithm comes with a set of assumptions, which we called its inductive bias, and each one may hold true on a different subset of the data. So we do not want to "put all our eggs in the same basket" but use them all.

For best accuracy, the different models we combine should be good by themselves and at the same time diverse to best complement each other. This is just like in real life where the best committee is composed of people having different areas of expertise; it makes no sense to consult multiple people if they all have very similar educations or backgrounds.[4]

One way to get diversity is by having models look at different sources of information. We already saw this in biometrics where the different models take different characteristics—for example, face, fingerprints, and so on—as input, and in speech recognition where in addition to the acoustic speech signal we also keep track of the speaker's lip.

Today, most of our data is multimedia, and *multi-view* models can be used in a variety of contexts where we have different sensors providing different but complementary information. In image retrieval, in addition to the image itself, we may also have a text description or a set of tag words. Using both sources together leads to better retrieval performance. Our smart devices, such as smart watches and smartphones, are equipped with sensors, and their readings can be combined for the purpose of, for example, *activity recognition*.

Outlier Detection

Another application area of machine learning is *outlier detection*, where the aim this time is to find instances that do not obey the general rule—those are the exceptions that are informative in certain contexts. The idea is that typical instances share characteristics that can be simply stated, and instances that do not have them are atypical.

In *Anna Karenina*, Tolstoy writes, "All happy families resemble one another, but each unhappy family is unhappy in its own way." This holds true in many domains, and not only for the case of nineteenth-century Russian families. For example, in medical diagnosis, we can similarly say that all healthy people are alike and that there are different ways of being unhealthy—each one of them is one disease.

In such a case, the model covers the typical instances and then any instance that falls outside is an exception. An *outlier* is an instance that is very much different from other instances in the sample. An outlier may indicate an abnormal behavior of the system; for example, for a credit card transaction, it may indicate *fraud*; in an image, an outlier may indicate an *anomaly* requiring attention, for example, a tumor; in the case of network traffic, it may be an intrusion attempt by a hacker; in a health-care scenario, it may indicate a significant deviation from a patient's normal behavior. Outliers may also be recording errors (e.g., due to faulty sensors) that should be detected and discarded to get reliable statistics. An outlier may also be a novel, previously unseen but valid case, which is where the related term, *novelty detection*, comes into play. For example, it may be a new type of profitable customer, indicating a new niche in the market waiting to be exploited by the company.

Dimensionality Reduction

In any application, observed data attributes that we believe contain information are taken as inputs and are used for decision making. However, it may be the case that some of these features actually are not informative at all, and they can be discarded; for example, it may turn out that the color of a used car does not have a significant effect on its price. Or, it may be the case that two different attributes are correlated and say basically the same thing (e.g., the production year and mileage of a used car are highly correlated), so keeping one may be enough.

We are interested in *dimensionality reduction* in a separate preprocessing step for a number of reasons:

First, in most learning algorithms, both the complexity of the model and the training algorithm depend on the number of input attributes. Here, complexity is of two types: the time complexity, which is how much calculation we do, and the space complexity, which is how much memory we need. Decreasing the number of inputs always decreases both, but how much they decrease depends on the particular model and the learning algorithm.

Second, when an input is deemed unnecessary, we save the cost of measuring it. For example, in medical diagnosis, if it turns out that a certain test is not needed, we do not do it, thereby eliminating both the monetary cost and the patient discomfort.

Third, simpler models are more robust on small data sets; that is, they can be trained with fewer data; or when trained with the same amount of data, they have smaller variance in their response, which indicates lower uncertainty.

Fourth, when data can be explained with fewer features, we have a simpler model that is easier to interpret.

Fifth, when data can be represented in few (e.g, two) dimensions, it can be plotted and analyzed visually, for structure and outliers, which again helps facilitate knowledge extraction from data. A plot is worth a thousand dots, and if we can find a good way to display the data, our visual cortex can do the rest, without any need for model fitting calculation.

There are basically two ways to achieve dimensionality reduction: feature selection and feature extraction. In *feature selection*, we keep the important features and discard the unimportant ones. It is basically a process of subset selection where we want to choose the smallest subset of the set of input attributes leading to maximum performance. The most widely used method for feature selection is the *wrapper* approach, where we iteratively add features until there is no further improvement. The feature selector is "wrapped around" the basic classifier or regressor that is trained and tested with each subset.

In *feature extraction*, we define new features that are calculated from the original features. These newly calculated features are fewer in number but still preserve the information in the original features. Those few synthesized features explain the data better than any of the original attributes, and sometimes they may be interpreted as hidden or abstract concepts.

In projection methods, each new feature is a linear combination of the original features; one such method is *principal component analysis* where we find new features that preserve the maximum amount of variance of the data. If the variance is large, the data has large spread making the differences between the instances most apparent, whereas if the variance is small, we lose the differences between data instances. The other method, *linear discriminant analysis* is a form of supervised feature extraction where the aim is to find new features that maximize the separation between classes.

Whether one should use feature selection or extraction depends on the application and the granularity of the features. If we are doing credit scoring and have features such as customer age, income, profession, and so on, feature selection makes sense. For each feature, we can say whether it is informative or not by itself. But a feature projection does not make sense: what does a linear combination (weighted sum) of age, income, and profession mean? On the other hand, if we are doing face recognition and the inputs are pixels, feature selection does not make sense—an individual pixel by itself does not carry discriminative information. It makes more sense to look at particular *combinations* of pixels in defining a face, as is done by feature extraction; for example, a region in image may define a particular type of eye or nose.

Nonlinear dimensionality reduction methods go beyond a linear combination and can find better features; this is one of the hottest topics in machine learning. The ideal feature set best represents the (classification or regression) information in the data set using the fewest numbers, and it is a process of encoding. It may also be considered as a process of abstraction because these new features can correspond to higher-level features representing the data in a more concise manner. In chapter 4, we will discuss autoencoder networks and deep learning where this type of nonlinear feature extraction is learned in artificial neural networks.

Decision Trees

Previously we discussed if-then rules and one way to learn such rules is by decision trees. The *decision tree* is one of the oldest methods in machine learning and though simple in both training and prediction, it is accurate in many domains. Trees use the famous "divide and conquer" strategy popular since Caesar where we divide a complex task—for example, governing Gaul—into simpler, regional tasks. Trees are used in computer science frequently for the same reason, namely to decrease complexity, in all sorts of applications.

Earlier we covered nonparametric estimation where, as you will remember, the main idea is to find a subset of the neighboring training examples that are most similar to the new query. In k-nearest-neighbor algorithms, we do this by storing all the training data in memory, calculating one by one the similarity between the new test query and all the training instances, and choosing the k most similar ones. This is rather a complex calculation when the training data is large, and it may be infeasible when the data is big.

The decision tree finds the most similar training instances by a sequence of tests on different input attributes. The tree is composed of decision nodes and leaves; starting from the root, each decision node applies a splitting test to the input and depending on the outcome, we take one

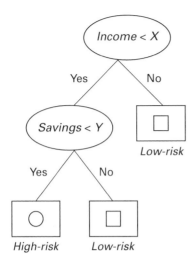

Figure 6   A decision tree separating low- and high-risk customers. This tree implements the discriminant shown in figure 3.

of the branches. When we get to a leaf, the search stops and we understand that we have found the most similar training instances, and we interpolate from those (see figure 6).

Each path from the root to a leaf corresponds to a conjunction of test conditions in the decision nodes on the path and such a path can be written as an if-then rule. That is one of the advantages of the decision tree: that a tree can be converted to a rule base of if-then rules and that those rules are easy to interpret. The tree is trained with a given training data where splits are placed to delimit regions that have the highest "purity," in the sense that each region contains instances that are similar in terms of their output.

Decision tree learning is nonparametric—we do not have a model where the structure is assumed a priori and we only update the parameters of that fixed structure; a decision tree grows as needed and its size depends on the complexity of the problem underlying the data; for a simple task, the tree is small, whereas a difficult task grows a large tree.

There are different decision tree models and learning algorithms depending on the splitting test used in the decision nodes and the interpolation done at the leaves; one very popular current approach is the *random forest*, where we train many decision trees on randomly chosen different subsets of the training data and we combine their predictions by taking a vote.

Trees are used successfully in various machine learning applications, and together with the linear model, the decision tree should be taken as one of the basic benchmark methods before any more complex learning algorithm is tried.

Active Learning

In learning, it is critical that the learner also knows what it knows and what it does not know. When a trained model makes a prediction, it is helpful if it can also indicate its certainty in that prediction.

One easy way to do this is by *resampling*.

Let us say that we want to be able to predict the price of a used car and we have a data set of 100 examples. Let us choose randomly 80 examples from that data and using those we train our first model. Then we can choose another random subset of 80 examples from the original data set and train a second model. The two data sets will be similar but not the same, and so the two fitted models will be similar in their predictions but not the same.

We can do this for example ten times and get ten models. Later on, when we are given a new car, we can give its mileage as input to all ten models and get ten predictions. They will be slightly different because each one has been trained on a slightly different data set. We can then use the average of those ten values as our point estimate; we can also sort those ten estimates from the smallest to the largest and define an interval, the so-called *confidence interval*, from the minimum to the maximum (in practice, it is better to discard the two extreme values at either end and use the interval from the second to the ninth). The size of that interval is a measure of our predictive uncertainty; if it is large we understand that our prediction is very much affected by slight changes in the data and hence is not too reliable.

The important point here is that data points are not created equal and our data may not be uniform all over the input space. For example, we may have a lot of cars in the training set with mileage less than 100K but not many cars with more.

In the case of estimation, in regions of the input space where we have a lot of data, we will have many training examples which means that all ten sets will have instances from there, which in turn implies that we would expect the ten models to make similar predictions in there and hence the confidence interval for any input there will be small. Where we have few data in the input space, it may be possible that there will not be examples in the training data of all of the ten models, and hence the predictions of the ten models can be expected to vary more and we will have a larger confidence interval.

So where the confidence interval is large, the uncertainty is understood to be large, and the model can actively ask the supervisor to provide training examples in there. This is called *active learning*. The model generates inquiries by synthesizing new inputs and asks for them to be supplied, rather like a student asking a question during a lecture.

For example, very early on in artificial intelligence, it was realized that in classification the most informative examples are those that lie closest to the current estimate of the class boundary: a *near miss* is an instance that looks very much like a positive example but is actually a negative example (Winston 1975).

A related research area in machine learning is called *computational learning theory*, where work is done to find theoretical bounds for learning algorithms that hold in general, independent of the particular learning task. For example, for a given model and learning algorithm, we may want to know the minimum number of training instances needed to guarantee at most a certain error with high enough probability—this is called *probably approximately correct learning* (Valiant 1984).

Learning to Rank

Ranking is an application area of machine learning that is different from regression or classification, and that is sort of between the two. In classification and regression, for each instance, we have a desired absolute value for the output; in ranking we train on pairs of instances and are asked to have the outputs for the two in the correct order.

Let us say we want to learn a *recommendation model* for movies. For this task, the input is composed of the attributes of the movie. The output is a numeric score that is a measure of how much we believe that a particular customer will enjoy a particular movie. To train such a model, we use past ratings by that customer. If we know that the customer liked movie A more than movie B in the past, we do our training such that for that customer the estimated score for A is indeed a higher value than the estimated score for B. Then, later on when we use that model to make a recommendation based on the highest scores, we expect to choose a movie that is more similar to A than to B.

There is no required numeric value for the score, as we have for the price of a used car, for example. The scores can be in any range as long as the ordering is correct. The training data is not given in terms of absolute values but in terms of such rank constraints (Liu 2011).

We can note here the advantage and difference of a ranker over a classifier or a regressor. If users rate the movies they have seen as enjoyed versus not enjoyed, this will be a two-class classification problem and a classifier can be used, but taste is nuanced, and a binary rating is hard to come by. On the other hand, if people rate their enjoyment of each movie on a scale of, say, from 1 to 10, this will be a regression problem, but such values are difficult to assign. It is more natural and easier for people to say of the two movies they watched which one they enjoyed more. After the ranker is trained with all such pairs, it is expected to generate numeric scores satisfying all these constraints.

Ranking has many applications. In search engines, we want to retrieve the most relevant documents when given a query. When we retrieve and display the current top ten candidates, if the user clicks the third one skipping the first two, we understand that the third should have been ranked higher than the first and the second. Such *click logs* are used to train rankers.

Bayesian Methods

In certain applications and with certain models, we may have some *prior belief* about the possible values of parameters. When we toss a coin, we expect it to be a fair coin or close to being fair, so we expect the probability of heads to be close to ½; in estimating the price of a car, we expect mileage to have a negative effect on the price. Bayesian methods allow us to take such prior beliefs into account in estimating the parameters.[5] The idea in *Bayesian estimation* is to use that prior knowledge together with the data to calculate a *posterior distribution* for the parameters. Let us see an example.

Assume we want to estimate the probability that people will click a certain link on our company's webpage; toward that end, we collect a sample of ten visitors and find that six of them have clicked on the link. The usual approach would be to say that the probability is 0.6 and use it for any further calculations and processing based on that value.

Now we can see that the actual probability could actually have been 0.5 and that it was just luck that we had seen six clicks in that particular sample of ten. It is even possible (though much less possible) that the actual probability is 0.1 and that a rare event occurred and we got six clicks. So what we can really get from the data is not just one value (0.6) but a list of possible values where for each we also have an additional value of how well the data supports it; this is what we mean by the posterior distribution. Once we have such a posterior distribution, we do further calculations using all (or a reasonable subset of) possible values and use their average, each weighted by how likely that value is.

The Bayesian approach is especially useful when the data set is small. If we see 60 clicks in 100 trials, the range of probable values around 0.6 will be much smaller.

Though the Bayesian approach is flexible and interesting, it has the disadvantage that except for simple scenarios under restrictive assumptions, the necessary calculation is too complex. One possibility is that of *approximation* where instead of the real posterior distribution that we cannot easily handle, we use one that is similar but manageable. Another possibility is *sampling* where instead of using the distribution itself, we generate representative instances from the distribution and make our inferences based on them. The popular methods for these—namely, *variational approximation* for the former, and *Markov chain Monte Carlo* (MCMC) *sampling* for the latter—are among important current research directions in machine learning.

The Bayesian approach allows us to incorporate our prior beliefs in training. One prior belief is that the underlying problem is smooth, which makes us prefer simpler models; remember our discussion of Occam's razor and the Kanizsa triangle from chapter 2. In *regularization*, we penalize complexity, and during training, in addition to maximizing our fit to the data, we also try to minimize the model complexity. While learning, we also get rid of those parameters that make the model unnecessarily complex and the output too variant. This implies a learning scheme that involves not only the adjustment of parameters but also changes to the model structure. Or we can go in the other direction and add complexity incrementally when we suspect we have a model that is too simple for the data.

The use of such *nonparametric* approaches in Bayesian estimation is especially interesting because we are no longer constrained by some parametric model class, but the model complexity also changes dynamically to match the complexity of the task in the data (Orbanz and Teh 2010). This implies a model of "infinite size," because it can be as complex as we want—it grows when it learns.

One model family that works quite well in many domains is the artificial neural network that is inspired from the human brain; in the next chapter, we will discuss how such networks are organized in layers and how such "deep" networks can be trained.

NEURAL NETWORKS AND DEEP LEARNING

Artificial Neural Networks

Our brains make us intelligent; we see or hear, learn and remember, plan and act thanks to our brains. In trying to build machines to have such abilities then, our immediate source of inspiration is the human brain, just as birds were the source of inspiration in our early attempts to fly. What we would like to do is to look at how the brain works and try to come up with an understanding of how it does what it does. But we want to have an explanation that is independent of the particular implementation details—this is what we called the *computational theory* when we discussed levels of analysis in chapter 1. If we can extract such an abstract, mathematical description, we can later implement it with what we have at our disposal as engineers— for example, in silicon and running on electricity.

Early attempts to build flying machines failed until we understood the theory of aerodynamics; only then we could build airplanes. Today, we see birds and airplanes as two different ways of flying—we call them airplanes now, not artificial birds, and they can do more than birds can; they cover longer distances and carry passengers or cargo. The idea is to accomplish the same for intelligence, and we start by getting inspired by the brain.

The human brain is composed of a very large number of processing units, called *neurons*, and each neuron is connected to a large number of other neurons through connections called *synapses*. Neurons operate in parallel and transfer information among themselves over these synapses. It is believed that the processing is done by the neurons and memory is in the synapses, that is, in the way the neurons are connected and influence each other.

Research on *neural networks* as models for analog computation—neuron outputs are not discrete, 0 or 1, but when they are activated they fire at a frequency which is a continuous value—started as early as research on digital computation (McCulloch and Pitts 1943) but, after the quick success and widespread use of digital computers, went largely unnoticed for a long time.

In the 1960s, the *perceptron* model was proposed as a model for pattern recognition (Rosenblatt 1962). It is a network composed of artificial neurons and synaptic connections, where each neuron has an activation value, and a connection from neuron A to neuron B has a weight that defines the effect of A on B. If the synapse is excitatory, when A is active it also tries to activate B; if the synapse is inhibitory, when A is active it tries to suppress B.

During operation, each neuron sums up the activations from all the neurons that make a synapse with it, weighted by their synaptic weights, and if the total activation is larger than a threshold value, the neuron "fires" and its output corresponds to the value of this activation; otherwise the neuron is silent. If the neuron fires, it sends its activation value in turn down to all the neurons with which it makes a synapse (see figure 7).

The perceptron basically calculates a weighted sum before making a decision, and this can be seen as one way of implementing a variant of the linear model we discussed earlier. Such neurons can be organized as layers where all the neurons in a layer take input from all the neurons in the previous layer and calculate their value in parallel, and these values together are fed to all the neurons in the layer that follows—this is called a *multilayer perceptron*.

Some of the neurons are sensory neurons and take their values from the environment, for example, from the sensed image, similar to the receptors in the retina. These then are given to other neurons that do some more processing over them in successive layers as activation propagates over the network. Finally, there are the output

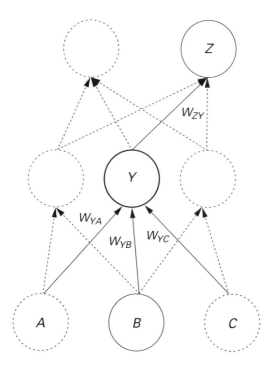

Figure 7 An example of a neural network composed of neurons and synaptic connections between them. Neuron Y takes its inputs from neurons A, B, and C. The connection from A to Y has weight W_{YA} that determines the effect of A on Y. Y calculates its total activation by summing the effect of its inputs weighted by their corresponding connection weights. If this is large enough, Y fires and sends its value to the neurons after it—for example, Z—through the connection with weight W_{ZY}.

neurons that make the final decision and carry out the actions through actuators—for example, to move an arm, utter a word, and so on.

Neural Network Learning Algorithms

In a neural network, learning algorithms adjust the connection weights between neurons. An early algorithm was proposed by Hebb (1949) and is known as the *Hebbian learning* rule: the weight between two neurons gets reinforced if the two are active at the same time—the synaptic weight effectively learns the correlation between the two neurons.

Let us say we have one neuron that checks whether there is a circle in the visual field and another neuron that checks whether there is the digit six, '6', in the visual field. Whenever we see a six—or are told that it is a six when we are learning to read—we also see a circle, so the connection between them is reinforced, but the connection between the circle neuron and, say, the neuron for digit seven, '7', is not reinforced because when we see one, we do not have the other. So the next time we see a circle in the visual field, this will increase the activation of the neuron for the digit six but will diminish the activation of the neuron for the digit seven, making six a more likely hypothesis than seven.

In some applications, certain neurons in the network are explicitly designated as input units and certain of them as output units. We have a training set that contains a sample of inputs and their corresponding correct output values, as specified by a supervisor—for example, in estimating the price of a used car, we have the car attributes as the input and their prices as the output. In this case of supervised learning, we clamp the input units to the input values in the training set, let the activity propagate through the network depending on the weights and the network structure, and then we look at the values calculated at the output units.

We define an *error function* as the sum of the differences between the actual outputs the network estimates for an input and their required values specified by the supervisor in the training set; and in neural network training, for each training example, we update the connection weights slightly, in such a way as to decrease the error for that instance. Decreasing the error implies that the next time we see the same or similar input, estimated outputs will be slightly closer to their correct values. *Theoretically speaking*, this is nothing but the good old regression we discussed in chapter 2, except that here the model is *implemented* as a neural network of neurons and connections.

This is one important characteristic of neural network learning algorithms, namely that they can learn *online*, by doing small updates on the connection weights as we see training instances one at a time. In *batch* learning, we have the whole data set and do training all at once using the whole data. A popular approach today involves *mini batches*, where we use small sets of instances in each update.

Currently with data sets getting larger, online learning is attractive because it does not require the collection and the storage of the whole data; we can just learn by using one example or a few examples at a time in a streaming data scenario. Furthermore, if the underlying characteristics of the data change slowly—as they generally do—online learning can adapt seamlessly, without needing to stop, collect new data, and retrain.

What a Perceptron Can and Cannot Do

Though the perceptron was successful in many tasks—remember that the linear model works reasonably well in many domains—there are certain tasks that cannot be implemented by a perceptron (Minsky and Papert 1969). The most famous of these is the exclusive OR (XOR) problem:

In logic, there are two types of OR, the inclusive OR and the exclusive OR. In everyday speech, when we say,

"To go to the airport, I will take the bus or the train," what we mean is the exclusive OR. There are two cases and only one of them can be true at one time. To represent the inclusive OR, we use the construct "and/or," as in "This fall, I will take Math 101 and/or Phys 101." In other words, I will take Math 101, Phys 101, or both.

Though the inclusive OR can be implemented by a perceptron, the exclusive OR cannot. It is not difficult to see why: if you have two cases, for example, the bus and the train, and if you want either to be enough, you need to give each of them a weight larger than the threshold so that the neuron fires when any one of them is true. But then when both of them are true, the overall activation will be twice as high and cannot be less than the threshold.

Though it was known at that time that tasks like XOR *can* be implemented using multiple layers of perceptrons, it was not known how to train such networks; and the fact that the perceptron cannot implement a task as straightforward as XOR—which can easily be implemented by a few (digital) logic gates—led to disappointment and the abandonment of neural network research for a long time, except for a few places around the world. It was only in the mid-1980s when the *backpropagation* algorithm was proposed to train multilayer perceptrons—the idea had been around since the 1960s and 1970s but had gone largely unnoticed—that interest in it was revived (Rumelhart, Hinton, and Williams 1986).

Recurrent Networks for Learning Time

Not all artificial neural networks are feedforward; there are also *recurrent networks* where in addition to connections between layers, neurons also have connections to neurons in the same layer (including themselves), or even to neurons back to the layers that precede them. Each activation calculation causes a certain delay so the *recurrent connections* act as a *short-term memory* for contextual information and let the network remember the past.

Let us say that input neuron A is connected to neuron X and that there is also a recurrent connection from X to itself (see figure 8). The effect of this connection is that at time t, the value of X will depend on input A at time t and will also depend on the value of X at time $t - 1$ because of the recurrent connection from X to itself. In the next time step, X at time $t + 1$ will depend on input A at time $t + 1$ and also on X at time t (previously calculated using A at time t and X at time $t - 1$), and so on. In this way, the value of X at any time will have depended on all the inputs seen until then.

If we define the state of a network as the collection of the values of all the neurons at a certain time, recurrent connections allow the current state to depend not only on the current input but also on the network state in the previous time steps calculated from the previous inputs. So, for example, if we are seeing a sentence one word at

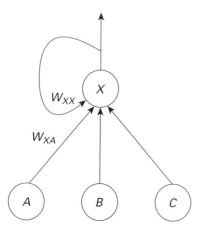

Figure 8 The recurrent connection acts as a short-term memory. The value of X depends not only on its immediate inputs A, B, and C, but also on its value in the previous time step and the weight W_{XX} of the recurrent connection.

a time, the recurrence allows the previous words in the sentence to be kept in this short-term memory in a condensed and abstract form and hence taken into account while processing the current word. The architecture of the network and the way recurrent connections are set define how far back and in what way the past influences the current output.

Recurrent neural networks are used in many tasks where the time dimension is important, as in speech or language processing, where what we would like to recognize are sequences. In a translation of text from one

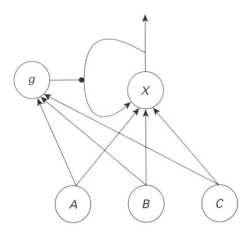

Figure 9 The "forget gate" g is another unit that sees the input A, B, and C and its output decides whether the activation passes or not through its associated recurrent connection. If g is 1, the gate is closed and the past is taken into account; if g is 0, the recurrent connection is cut and the past value of X does not have any effect on the next value, that is, the unit has effectively forgotten the past.

language to another, not only the seen input but also the generated output is a sequence.

More complex types of recurrent connections are also possible. In a long short-term memory (LSTM) unit (Hochreiter and Schmidhuber 1997), there is a "forget gate" where this type of effect can be turned on or off (see figure 9). So depending on where and how the gate is turned on and off, the network can be selective as to what to remember from the past.

For example, compare the following two sentences:

"The man entered the room, looked around, and took off his jacket."

"The woman entered the room, looked around, and took off her jacket."

Whether the possessive pronoun is "his" or "her" depends on the gender of the subject, that is, whether the person entering the room is a man or woman. So in generating such a sentence one word at a time (e.g., when generating a translation), the network should store that gender information when it processes the subject word at the beginning of the sentence and keep it intact (unaffected by the following words that are irrelevant for this purpose) until it generates the correct pronoun.

Connectionist Models in Cognitive Science

Artificial neural network models are known as *connectionist* or *parallel distributed processing* (PDP) models in cognitive psychology and cognitive science (Feldman and Ballard 1982; Rumelhart and McClelland and the PDP Research Group 1986). The idea is that neurons correspond to concepts and that the activation of a neuron corresponds to our current belief in the truth of that concept. Connections correspond to constraints or dependencies between concepts: A connection has a positive weight and is excitatory if the two concepts occur simultaneously—for example, between the neurons for circle and '6'—and has a negative weight and is inhibitory if the two concepts are mutually exclusive—for example, between the neurons for circle and '7'.

Neurons whose values are observed—for example, by sensing the environment—affect the neurons they are connected to, which in turn affect the neurons *they* are connected to, and so on. This activity propagation throughout the network results in a state of neuron outputs that satisfies the constraints defined by the connections.

The basic idea in connectionist models is that intelligence is an *emergent property* and high-level tasks, such as recognition or association between patterns, arise automatically as a result of this activity propagation by the rather elemental operations of interconnected simple processing units. Similarly, learning is done at the connection level through simple operations, for instance, according to the Hebbian rule, without any need for a higher-level programmer.

Connectionist networks care about biological plausibility but are still abstract models of the brain; for example, it is very unlikely that there is actually a neuron for every concept in the brain—this is the *grandmother cell* theory,

Intelligence is an *emergent property* and tasks such as recognition arise automatically as a result of the propagation of activity over a network of simple processing units.

which states that I have a neuron in my brain that is activated only when I see or think of my grandmother—that is a *local representation*. It is known that neurons die and new neurons are born in the brain, so it makes more sense to believe that the concepts have a *distributed representation* on a cluster of neurons, with enough redundancy for concepts to survive despite physical changes in the underlying neuronal structure.

Neural Networks as a Paradigm for Parallel Processing

Since the 1980s, computer systems with thousands of processors have been commercially available. The software for such parallel architectures, however, has not advanced as quickly as hardware. The reason for this is that almost all our theory of computation up to that point was based on serial, single-processor machines. We are not able to use the parallel machines in their full capacity because we cannot program them efficiently.

There are mainly two paradigms for *parallel processing*. In *single instruction, multiple data* (SIMD) machines, all processors execute the same instruction but on different pieces of data. In *multiple instruction, multiple data* (MIMD) machines, different processors may execute different instructions on different data. SIMD machines are easier to program because there is only one program to write.

However, problems rarely have such a regular structure that they can be parallelized over a SIMD machine. MIMD machines are more general, but it is not an easy task to write separate programs for all the individual processors; additional problems arise that are related to synchronization, data transfer between processors, and so forth. SIMD machines are also easier to build, and machines with more processors can be constructed if they are SIMD. In MIMD machines, processors are more complex, and a more complex communication network must be constructed for the processors to exchange data arbitrarily.

Assume now that we can have machines where processors are a little bit more complex than SIMD processors but not as complex as MIMD processors. Assume that we have simple processors with a small amount of local memory where some parameters can be stored. Each processor implements a fixed function and executes the same instructions as SIMD processors; but by loading different values into its local memory, each processor can be doing different things and the whole operation can be distributed over such processors. We will then have what we can call *neural instruction, multiple data* (NIMD) machines, where each processor corresponds to a neuron, local parameters correspond to its synaptic weights, and the whole structure is a neural network. If the function implemented in each processor is simple and if the local memory is small, then many such processors can be fit on a single chip.

The problem now is to distribute a task over a network of such processors and to determine the local parameter values. This is where learning comes into play: We do not need to program such machines and determine the parameter values ourselves if such machines can learn from examples.

Thus, artificial neural networks are a way to make use of the parallel hardware we can build with current technology and—thanks to learning—they need not be programmed. Therefore, we also save ourselves the effort of programming them.

Hierarchical Representations in Multiple Layers

Before, we mentioned that a single layer of perceptron cannot implement certain tasks, such as XOR, and that such limitations do not apply when there are multiple layers. Actually, it has been proven that the multilayer perceptron is a *universal approximator*, that is, it can approximate any function with desired accuracy given enough neurons— through training it to accomplish that is not always straightforward.

The perceptron algorithm can train only single-layer networks, but in the 1980s the *backpropagation algorithm* was invented to train multilayer perceptrons, and this caused a flurry of applications in various domains

Artificial neural networks are a way to make use of the parallel hardware we can build with current technology and—thanks to learning—they need not be programmed.

significantly accelerating neural network research in many fields, from cognitive science to computer science and engineering.

The multilayer network is intuitive because it corresponds to layers of operation where we start from the raw input and incrementally perform a more complicated transformation, until we get to the output.

For example, in image recognition, we have image pixels as the basic input and as input to the first layer. The neurons in the next layer combine these to detect basic image descriptors such as strokes and edges of different orientations. A later layer combines these to form longer lines, arcs, and corners. Layers that follow combine them to learn more complex shapes such as circles, squares, and so on. These in turn are combined with some more layers of processing to represent the objects we want to learn, such as faces or handwritten characters.

Each neuron in a layer defines a more complex feature in terms of the simpler patterns detected in the layer below it. These intermediate feature-detecting units are called *hidden units* because they correspond to hidden attributes not directly observed but are defined in terms of what is observed. These successive layers of hidden units correspond to increasing layers of abstraction, where we start from raw data such as pixels and end up in abstract concepts such as a digit or a face.

It is interesting to note that a similar mechanism seems to be operating in the visual cortex. In their experiments on cats, Hubel and Wiesel, who were later awarded the 1981 Nobel Prize for their work on visual neurophysiology, have shown that there are *simple cells* that respond to lines of particular orientations in particular positions in the visual field, and these in turn feed to *complex* and *hypercomplex cells* for detecting more complicated shapes (Hubel 1995)—though not much is known about what happens in later layers.

Imposing such a structure on the network implies making assumptions, such as dependencies, about the input. For example, in vision we know that nearby pixels are correlated and there are local features like edges and corners. Any object, such as a handwritten digit, may be defined as a combination of such primitives. We know that because the visual scene changes smoothly, nearby pixels tend to belong to the same object, and where there is sudden change—an edge—is informative because it is rare.

Similarly, in speech, locality is in time, and inputs close in time can be grouped as speech primitives. By combining these primitives, longer utterances, namely speech phonemes, can be defined. They in turn can be combined to define words, and these in turn can be combined as sentences.

In such cases, when designing the connections between layers, units are not connected to all of the input units because not all inputs are dependent. Instead, we define units that define a window over the input space and are connected to only a small *local* subset of the inputs. This decreases the number of connections and therefore the number of parameters to be learned. Such a structure is called a *convolutional neural network* where the operation of each unit is considered to be a convolution—that is, a matching—of its input with its weight (Le Cun et al. 1989). An earlier similar architecture is the *neocognitron* (Fukushima 1980).

The idea is to repeat this in successive layers where each layer is connected to a small number of local units below. Each layer of *feature extractors* checks for slightly more complicated features by combining the features below in a slightly larger part of the input space, until we get to the output layer that looks at the whole input. Feature extraction also implements *dimensionality reduction* because although the raw attributes that we observe may be many in number, the important hidden features that we extract from data and that we use to calculate the output are generally much fewer.

This multilayered network is an example of a *hierarchical cone* where features get more complex, abstract, and fewer in number as we go up the network until we get to classes (see figure 10).

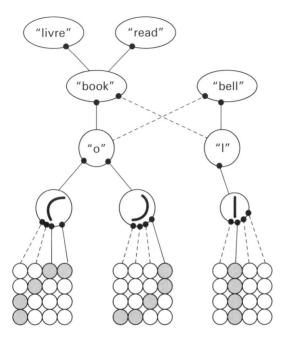

Figure 10 A very simplified example of hierarchical processing. At the lowest level are pixels, and they are combined to define primitives such as arcs and line segments. The next layer combines them to define letters, and the next combines them to define words. The representation becomes more abstract as we go up. Continuous lines denote positive (excitatory) connections, and dashed lines denote negative (inhibitory) connections. The letter o exists in "book" but not in "bell." At higher levels, activity may propagate using more abstract relationships such as the relationship between "book" and "read," and in a multilingual context, between "book" and "livre," the French word for book.

Deep Learning

In computer vision in the last half century, significant research has been done to find the best features for accurate classification, and many different image filters, transforms, and convolutions have been proposed to implement such feature extractors manually.

Though these approaches have had some success, learning algorithms are achieving higher accuracy recently with big data and powerful computers. With few assumptions and little manual interference, structures similar to the hierarchical cone are being automatically learned from large amounts of data. These learning approaches are especially interesting in that, because they learn, they are not fixed for any specific task, and they can be used in a variety of applications. They learn both the hidden feature extractors and also how they are best combined to define the output.

This is the idea behind *deep neural networks* where, starting from the raw input, each hidden layer combines the values in its preceding layer and learns more complicated functions of the input. The fact that the hidden unit values are not 0 or 1 but continuous allows a finer and graded representation of similar inputs (For example, if what we see in a small patch looks like a corner but is not exactly, the output of the corner-detecting hidden unit in that region will be, say, 0.7). Successive layers correspond to more abstract representations until we get to the final layer where the outputs are learned in terms of these most abstract concepts.

We saw an example of this in the convolutional neural network where starting from pixels, we get to edges, and then to corners, and so on, until we get to a digit. In such a network, some user knowledge is necessary to define the connectivity and the overall architecture. Consider a face recognizer network where inputs are the image pixels. If each hidden unit is connected to *all* the pixels, the network has no knowledge that the inputs are face images or even that the input is two-dimensional—the input is just a set of values. Using a convolutional network where hidden units are fed with localized two-dimensional patches is a way to feed this locality information such that correct abstractions can be learned.

In *deep learning*, the idea is to learn feature levels of increasing abstraction with minimum human contribution (Goodfellow et al. 2016; LeCun, Bengio, and Hinton 2015; Schmidhuber 2015). This is because in most applications, we do not know what structure there is in the input, especially as we go up and the corresponding concepts become "hidden." So, any sort of dependency should be automatically discovered during training from a large sample of examples. It is this extraction of hidden dependencies, or patterns, or regularities from data that allows abstraction and learning general descriptions.

In chapter 2 when we discussed the example of fitting a model to a sequence of numbers, we saw that as the sequence gets more complex, we need more flexible models to be able to make a fit. Learning basically is a process of matching the complexity of the learner model to that of the task underlying the data. The representational capability of a neural network depends on its number of layers and units in each layer, so, as the task that we want to learn gets complex, we need deeper networks with more layers and units.

Training a network with multiple hidden layers is difficult and slow because the error at the output needs to be propagated back to update the weights in all the preceding layers, and there is interference when there are many parameters. In a convolutional network, each unit is fed to only a small subset of the units before and feeds to only a small subset of units after, so there is less interference and training can be done faster.

Deep learning methods are attractive mainly because they need less manual help. We do not need to craft the right features or the suitable transformations. Once we have data—and today we have "big" data—and sufficient computation available—and today we have data centers with thousands of processors—we just wait and let the learning algorithm discover all that is necessary by itself.

Another important factor that fueled deep learning research in recent years is the availability of software libraries that allow coding deep neural networks very easily. Those libraries can also efficiently utilize parallel hardware and thus permit testing different network architectures very quickly.

The idea of multiple layers of increasing abstraction that underlies deep learning is intuitive. Not only in vision—in handwritten digits or face images—but also in many applications we can think of such layers of abstraction. Discovering these abstract representations is useful, not only for prediction but also because abstraction allows a better description and understanding of the problem.

Learning Hidden Representations

A special type of multilayer network is the *autoencoder*, where the desired output is set to be equal to the input, and the network has a hourglass shape with fewer hidden units in the intermediate layers than there are in the input and output. Such a network is composed of two parts where the first part, from the input to the hidden layer, implements an *encoder* stage where a high-dimensional input is compressed to be represented by the values of the fewer hidden units. The second part, from the hidden layer to the output, implements a *decoder* stage that takes that low-dimensional representation in the hidden layer and

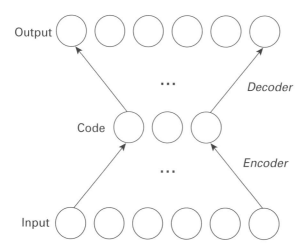

Output

Decoder

Code

Encoder

Input

Figure 11 The autoencoder is a neural network where there are fewer hidden units than input units, and the output is set to be equal to the input. The encoder needs to learn to generate a short, compressed "code" in its hidden layer that should be sufficient for the decoder to be able to reconstruct the input back at the output.

reconstructs the higher dimensional input back again at the output (see figure 11).

For the network to be able to reconstruct the input at its output units, those few hidden units that act as a bottleneck should be able to extract the features that preserve information maximally. The autoencoder is unsupervised; those hidden units learn themselves without any supervision to find a good encoding of the input, a short, compressed description, extracting the most important features and ignoring what is irrelevant, namely, noise.

Researchers have proposed extensions and different uses of the basic autoencoder. For example, the autoencoder shown in figure 11 has only one layer of connections between the input and the hidden units, but in practice in a *deep* autoencoder, there may be multiple layers in the encoder (with their structure mirrored in the decoder) to be able to learn more abstract hidden representations; for example, if we have images as the input, the first few layers of the encoder are typically convolutional.

An interesting variant is the *noisy autoencoder* where the aim is to learn a hidden representation that is robust to perturbations of the input. Let us say we have face images and we see that some people wear glasses that occlude their eyes, which can mess up recognition. What we do is we use an autoencoder where we take two images of the same person as input, one with and one without glasses, and for both, we set the image without glasses as the desired output. To be able to generate the same output for both, the encoder should learn to generate the same code for both, meaning that it should learn to discard the occluding effect of the glasses. Such a representation generated by the encoder can then be given to a face recognizer which can make decisions despite glasses.

This can be done with any perturbation of the input that we want to get invariance to, for example, small rotations or translations. That is, if you have an input *x* and its perturbed version x^* and for both, we set x as the desired output, the network, assuming that it is big enough and that it sees enough training examples, learns to generate a hidden representation that is invariant to that type of perturbation.

The general idea is that if we have two different inputs for which we set the same desired output, the encoder will be forced to learn to generate the same, or very similar codes for the two. One application of this is in learning word representations in natural language processing. This is a topic where the need for good feature extractors, that is, good hidden representations, is most apparent. Researchers have worked on predefined databases, called *ontologies*, for representing relationships between words in a language and such databases work with some success; but again it turned out that the best way is to learn such relationships from a lot of data.

In the *word2vec* network which has the same architecture as an autoencoder, the output is a word and the input is a word in its context, that is, one of the words that are nearby in the same sentence (Mikolov et al. 2013). The result of such a training is that if two words appear in the same or very similar contexts, the encoder will generate similar codes for them.

For example, consider the following sentence:

"Visitors to Paris will enjoy its numerous museums."

It is highly possible that if we go through a very large corpus of sentences, we will also find very similar sentences but with "Berlin" instead of "Paris," or "Rome," and so on. We will have many sentences about cities, which are all almost the same, the only difference being the name of the city. This will make codes (encoder outputs) for all these cities to be similar because the same context words need to be decoded from them.

Now consider this sentence:

"The French foreign minister has returned to Paris."

Again, we will have many similar sentences with "German" instead of "French" and this time "Berlin" instead of "Paris." This will cause similar codes to be generated for "French" and "German," but also the representational relationship between "German" and "Berlin" will be the same as the one between "French" and "Paris." This leads to what is called *vector algebra*: Because codes are numbers, we can do arithmetic on them.

Let us say vec("Paris") denotes the learned code for "Paris" (see figure 12). After training, we expect vec("Paris") and vec("Berlin") to be nearby, and also vec("French") and vec("German") to be nearby but in some other part of the code space, but we expect also the relationships to be similar. That is, we expect vec("German") – vec("Berlin") to be very similar to vec("French") – vec("Paris"), so much

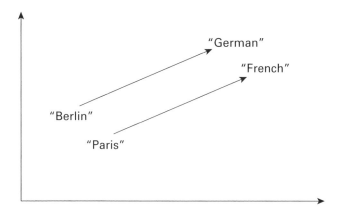

Figure 12   In the code space learned by word2vec, cities form one cluster and the adjectives denoting the country of origin form another cluster in some other part of the space. The relative positions are also expected to be very similar so that we can do vector algebra: We expect vec("German") −vec("Berlin") + vec("Paris") to be close to vec("French").

so that if we calculate vec("German") − vec("Berlin") + vec("Paris"), we will get a code very close to vec("French").

End-to-End Learning

In many applications, the processing can be viewed as an encoder-decoder structure that we have with the autoencoder discussed earlier.

Consider machine translation. Starting with an English sentence, in multiple levels of processing and abstraction that are learned automatically from a very large English corpus to code the lexical, syntactic, and semantic rules of the English language, we would get to the most abstract representation. Now consider the same sentence in French. The levels of processing learned this time from a French corpus would be different, but if the two sentences mean the same, at the most abstract, language-independent level, they should have very similar representations.

Language understanding is a process of encoding where from a given sentence, we extract this high-level abstract representation, and language generation is a process of decoding, where we synthesize a natural language sentence from such a high-level representation. In translation, we encode in the source language and decode in the target language. In a dialogue system, we first encode the question to an abstract level and process it to form a response in the abstract level, which we then decode as the response sentence.

The advantage of such a structure is that learning is end to end. We only specify the input to the encoder and the desired output at the very end of the decoder; it is enough to just provide a very large data set of input and output pairs, and any transformation needed in between is automatically learned by the many hidden layers of a deep network. Learning not only adjusts the parameters of the encoder and the decoder but also specifies the intermediate code between the two modules; this intermediate code is the representation of the input best suited to generate the corresponding output.

There are many interesting applications of deep neural networks trained end to end, and they typically have this structure. There is the early module that analyzes the input and transforms it into an intermediate representation and the later module learns to synthesize the correct output from that intermediate representation.

One example is the show-and-tell deep architecture that learns to generate captions for images (Vinyals et al. 2014). The encoder is a convolutional network that takes an image and analyzes it in its many levels to generate a code that summarizes the content of the image. The decoder is a recurrent network that generates the caption one word at a time from this code. The whole structure is trained end to end, from a large set of example pairs of images and manually provided captions.

Another example is the deep neural network that learns to play Atari games (Mnih et al. 2015). The input is the game screen and the output is the correct joystick action. The network has early convolutional layers that analyzes the image to extract the best features for deciding on the right action and the later, fully connected layers generate the action based on those.

It is also possible to use a network trained on one task to help solve another task, this is called *transfer learning*.

Let us say we want to do face recognition, but our data set is relatively small. Let us assume that we already have a very deep convolutional network trained on some very large image data set. What we do is we take the first few layers of that larger network and copy them to act as early layers in our face recognition network. If we expect the basic features necessary for the large problem to be also useful for face recognition, it makes sense to do this. We only train the later layers of the face recognition network, which means that there will be fewer parameters, and we can train those using a smaller data set.

Generative Adversarial Networks

The *generative adversarial network* (GAN) is actually composed of two networks, a generator G and a discriminator D (Goodfellow et al. 2014). The aim is to learn a generator; D is only there to train G. Both G and D are typically deep neural networks, but GAN is a general strategy for training that is independent of how the two learners are implemented.

The task of a generator is different from that of a regressor or a classifier. Let us say we have a training set of faces. What we would like to do is to learn the structure of faces from this data so well that we can generate a new face when we want. The output will be an image that looks like a face, it will have hair on top, eyes and nose suitably placed with respect to each other, the face should be symmetric, and so on, and all those constraints are to be learned from the data. Again, we want generalization, we do not want to generate a face already in the training set but we want the new image to be the face of a person outside of the training set; it will be the face of a person who does not even exist. Or let us say we have a training set of Bach chorales and we want to train a generator from those so that it can spit out a new chorale when we want.

In GAN, the generator takes a random input and transforms it into an instance, for example, a face image or a chorale (see figure 13). Different random inputs generate different instances. The input z has some predefined distribution and G maps each z to a candidate x. What G does is that it takes the distribution of z as input and stretches, translates, rotates, etc. it in its many successive layers such that its output looks as much as possible to the distribution of x.

Those instances that are generated by G are labelled as "fake." We also have a training set of actual faces/chorales and they are labelled as "true." D is a two-class classifier that is trained to separate fakes from true instances. G is like a forger that paints fake Rembrandts and D is like an art expert who is good at spotting fake Rembrandts.

G is trained to generate fakes that will be classified as true by D. The two are trained together; as D gets better, G

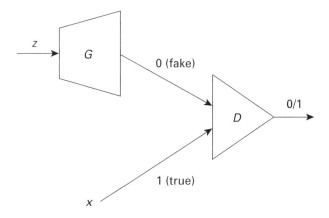

Figure 13 In the generative adversarial network, G is the generator that transforms a random z to a candidate x, but because it is generated it is called a "fake" instance. We also have the "true" x that are valid instances drawn from a training set. D is trained to separate fakes from true instances as well as possible; G is trained to generate fakes so well that D will classify them as true.

will use that information to understand what D classifies as true and will learn to generate fakes similar to those. D will in turn learn to tell these apart which in turn will force G to generate even better fakes, and so on.

GANs are one of the most popular research topics in machine learning these days, with very impressive results.[1] One major problem with GAN is that because there are two networks, training is more difficult; another problem is that the goodness of a generator is still largely evaluated manually, so GAN currently is mostly used in image generation tasks where such evaluation is done manually.

Until now, we have talked about supervised learning where there is an input and an output, and the aim is to learn the mapping from the input to the output. In the next chapter, we will discuss unsupervised learning where there is no explicit output, and the aim is to learn the regularity in the input space, to learn what type of things happen frequently.

LEARNING CLUSTERS AND RECOMMENDATIONS

Finding Groups in Data

Previously we covered supervised learning where there is an input and an output—for example, car attributes and price—and the aim is to learn a mapping from the input to the output. A supervisor provides the correct values, and the parameters of a model are updated so that its output gets as close as possible to these desired outputs.

We are now going to discuss *unsupervised learning*, where there is no predefined output, and hence no such supervisor; we have only the input data. The aim in unsupervised learning is to find the regularities in the input, to see what normally happens. There is a structure to the input space such that certain patterns occur more often than others, and we want to see what generally happens and what does not.

One method for unsupervised learning is *clustering*, where the aim is to find clusters or groupings of input; in statistics, these are called *mixture models*.

In the case of a company, the customer data contains demographic information, such as age, gender, zip code, and so on, as well as past transactions with the company. The company may want to see the distribution of the profile of its customers, to see what type of customers frequently occur. In such a case, a clustering model allocates customers similar in their attributes to the same group, providing the company with natural groupings of its customers; this is called *customer segmentation* (see figure 14). Once such groups are found, the company may decide strategies, for example, services and products, specific to different groups; this is known as *customer relationship management* (CRM).

Such a grouping also allows the company to identify those who are outliers, namely, those who are different from other customers, which may imply a niche in the market that can be further exploited by the company, or those customers who require further investigation, for example, churning customers.

We expect to see regularities and patterns repeated with minor variations in many different domains. Detecting them as primitives and ignoring the irrelevant variations is also a way of doing compression. For example, in an image, the input is made up of pixels, but we can

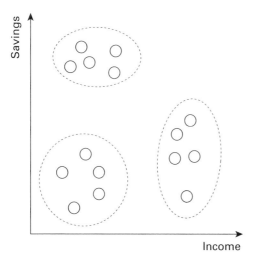

Figure 14 Clustering for customer segmentation. For each customer, shown
by a circle, we have the income and savings information. Here, we see that
there are three customer segments. Such a grouping allows us to understand
the characteristics of the different segments—for example, the segment
on the lower left is that of customers with low income and low savings—so
that we can define different interactions with each segment; this is called
customer relationship management.

identify regularities by analyzing repeated image patterns, such as, texture, objects, and so forth. This allows
a higher-level, simpler, and more useful description of the
scene and achieves better compression than compressing
at the pixel level. A scanned document page does not have
random on/off pixels but bitmap images of characters;
there is structure in the data, and we make use of this

redundancy by finding a shorter description of the data in terms of strokes of different orientations. Going further, if we can discover that those strokes combine in certain ways to make up characters, we can use just the code of a character, which is shorter than its image.

In *document clustering*, the aim is to group similar documents. For example, news reports can be subdivided into those related to politics, sports, fashion, arts, and so on. We can represent the document as a bag of words using a lexicon that reflects such document types, and then documents are grouped depending on the number of shared words. It is of course critical how the lexicon is chosen.

Unsupervised learning methods are also used in *bioinformatics*. DNA in our genome is the "blueprint of life" and is a sequence of bases, namely, A, G, C, and T. RNA is transcribed from DNA, and proteins are translated from RNA. Proteins are what the living body is and does. Just as DNA is a sequence of bases, a protein is a sequence of amino acids (as defined by bases). One application area of computer science in molecular biology is alignment, which is matching one sequence to another. This is a difficult string-matching problem because strings may be quite long, there are many template strings to match against, and there may be deletions, insertions, or substitutions.

Clustering is used in learning *motifs*, which are sequences of amino acids that occur repeatedly in proteins. Motifs are of interest because they may correspond to structural or functional elements within the sequences they characterize. The analogy is that if the amino acids are letters and proteins are sentences, motifs are like words, namely, a string of letters with a particular meaning occurring frequently in different sentences.

Clustering may be used as an *exploratory data analysis* technique where we identify groups naturally occurring in the data. We can then, for example, label those groups as classes and later on try to classify them. A company may cluster its customers and find segments, and then toward a certain aim—for example, churning—can label them and train a classifier to predict the behavior of new customers. But the important point is that there may be a cluster or clusters that no expert could have foreseen, and that is the power of unsupervised data-driven analysis.

Sometimes a class is made up of multiple groups. Consider the case of optical character recognition. There are two ways of writing the digit seven; the American version is '7', whereas the European version has a horizontal bar in the middle (to tell it apart from the European '1', which keeps the small stroke on top in handwriting). In such a case, when the sample contains examples from both continents, the class for seven should be represented as the union/disjunction/mixture of two groups.

A similar example occurs in speech recognition where the same word can be uttered in different ways, due to differences in pronunciation, accent, gender, age, and so on—"I say to-may-to, you say to-mah-to." Thus when there is not a single, universal way, all these different ways should be represented as equally valid alternatives to be statistically correct.

Clustering algorithms group instances in terms of their similarities calculated using their input representation, which is a list of input attributes, and the similarity between instances is measured by combining similarities in these attributes. In certain applications, we can define a similarity measure between instances directly, in terms of the original data structure, without explicitly generating such a list of attributes and calculating similarities over them.

Consider clustering Web pages. In addition to the text field, we can also use the similarity of meta (or header) information such as titles or keywords, or the number of common Web pages that link to or are linked from those two. This gives us a much better similarity measure than what is calculated using the bag of words representation on the text of the Web pages. Using a similarity measure that is better suited to the application—if one can be defined—leads to better clustering results; this is the basic idea in *spectral clustering*.

Such application-specific similarity representations are also popular in supervised learning applications typically grouped under the name *kernel function*. The *support vector machine* (Vapnik 1998) is one such learning algorithm used for both classification and regression.

It is also possible to do *hierarchical clustering*, where instead of a flat list of clusters, we generate a tree structure with clusters at different levels of granularity and clusters higher in the tree are subdivided into smaller clusters (see figure 15). We are familiar with such trees of clusters from studies in biology—most famously, the taxonomy by

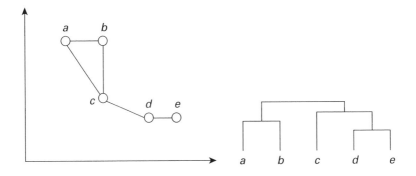

Figure 15 Example of hierarchical clustering. On the left, we have five instances, *a* to *e*, represented in two dimensions; these may for example be five customers and the two axes may be two attributes, such as income and savings. Closest instances are merged iteratively to define larger clusters and the structure can be visualized as a tree, as shown in the right. The advantage of such an approach is that we get different clustering solutions at different levels of granularity: At one extreme (where we have high tolerance to distance between instances), we have one cluster containing all five instances; at the other extreme (where we have very low tolerance), we have five clusters each containing one instance. One intermediate solution has three clusters, {*a, b*}, {*c*}, and {*d, e*}; this is what we get if our tolerance is less than the distance between *c* and *d*.

Linnaeus—or human languages. One explanation of the splitting up of clusters into smaller clusters is due to *phylogeny*, that is, to evolutionary changes—small mutations are gradually accumulated over time until a species (or a language) splits into two—but in other applications, the reason of similarity may be different.

The aim in clustering in particular, or unsupervised learning in general, is to find structure in the data. In the case of supervised learning (e.g., in classification), this structure is imposed by the supervisor who defines the different classes and labels the instances in the training data by these classes. This additional information provided by the supervisor is of course useful, but we should always make sure that it does not become a source of bias or impose artificial boundaries. There is also the risk that there is error in labeling, which is called "teacher noise."

Unsupervised learning is an important research area because unlabeled data is a lot easier and cheaper to find. For speech recognition, a talk radio station is a source of unlabeled speech data; spoken speech is not a random sequence of sounds, but we have particular sound sequences repeated frequently, which are the words in that language. The idea is to extract the basic characteristics from unlabeled data and learn what is typical, which can then later be labeled for different purposes. A baby spends their first few years looking around when they see things, objects, faces repeatedly under a variety of conditions,

The aim in clustering in particular, or unsupervised learning in general, is to find structure in the data, by extracting the basic characteristics and learning what is typical.

during which presumably they learn their basic feature extractors and how they typically combine to form objects. Later on, when that baby learns language, they learn the names for those.

Recommendation Systems

In chapter 1, we discussed *recommendation systems* for predicting customer behavior as an application of machine learning. Given a large data set of customer transactions, we can find *association rules* of the form, "People who buy X are also likely to buy Y." Such a rule implies that among the customers who buy X, a large percentage have also bought Y. So, if we find a customer who has bought X but has not bought Y, we can target them as a potential Y customer. X and Y can be products, authors, cities to be visited, videos to be watched, and so on; we see many examples of this type of recommendation every day, especially while surfing online.

Though this targeting approach is used frequently, and efficient algorithms have been proposed to learn such rules from very large data sets, interesting algorithms that make use of *generative models* are being proposed these days.

Remember that while constructing a generative model, we think about how we believe the data is generated. In customer behavior therefore, we consider the causes that affect this behavior. We know that people do not buy things at random. Their purchases depend on a number of factors, such as their household composition—that is, how many people they live with, their gender, ages—and their income, their taste (which in turn is a result of other factors such as the place of origin), and so on. Though some companies have loyalty cards and collect some of this information, in practice, most of these factors are not known, are hidden, and need to be inferred from the observed data.

Note, however, that even if we have some idea about what such factors, an overreliance on them can be misguided because they are often wrong or incomplete; there may also be factors that we cannot immediately think of or factors that are not as important as we think, which is why it is always best to learn (discover) them from data.

Extracting such hidden causes will build a much better model than trying to learn associations among products. For example, a hidden factor may be "baby at home," which will lead to the purchase of different items such as diapers, milk, baby formula, wipes, and so on. So instead of learning association rules between pairs or triples of these items, if we can estimate the hidden baby factor based on past purchases, this will trigger an estimation of whatever it is that has not been bought yet.

In practice, there are many such factors; each customer is affected (or defined) by a number of these, and each factor triggers a subset of the products. The factor values are not 0 or 1 but continuously valued, and this distributed representation provides a richness when it comes to representing customer instances.

This approach aims to find structure by decomposing data into two parts. The first one, the mapping between customers and factors, defines a customer in terms of the factors (with different weights). The second one, the mapping between factors and products, defines a factor in terms of the products (with different weights). In mathematics, we model data using matrices, which is why this approach is called *matrix decomposition*, or sometimes *tensor* decomposition, tensors being matrices with more than two dimensions.

Such a generative approach with hidden factors makes sense in many other applications. Let us take the case of movie recommendations (see figure 16). We have customers who have rented a number of movies and we have a score for each movie they watched, and from those we want to make a recommendation.

The first characteristic of this problem is that we have many customers and many movies, but the data is *sparse*. Every customer has watched only a small percentage of the movies, and most movies have been watched by only a small percentage of the customers. Based on these facts, the learning algorithm needs to be able to generalize and

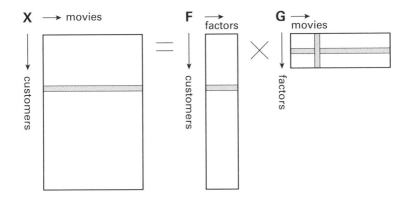

Figure 16 Matrix decomposition for movie recommendations. Each row of the data matrix X contains the scores given by one customer for the movies, most of which will be missing (because the customer hasn't watched that movie). It is factored into two matrices F and G where each row of F is one customer defined as a vector of factors and each row of G defines the effect of one factor over the movies; each column of G is one movie defined in terms of the factors. The number of factors is typically much smaller than the number of customers or movies; in other words, it is the number of factors that defines the complexity of the data, named the rank of the data matrix X.

predict successfully, even when new movies or new customers are added to the data.

In this case too, we can think of hidden factors, such as the age and gender of the customer, which makes certain genres, such as action, comedy, and so on, a more likely choice. Using decomposition, we can define each customer in terms of such factors (in different proportions), and each such factor triggers certain movies (with different probabilities). This is better—easier, cheaper—than trying to come up with rules between pairs of movies. Note again that such factors are not predefined but are automatically discovered during learning; they may not always be easy to interpret or assign a meaning to.

Another possible application area is *document categorization* (Blei 2012). Let us say we have a lot of documents, and each is written using a certain bag of words. Again the data is sparse; each document uses only a small number of words. Here, we can interpret hidden factors as topics. When a reporter writes a report, they want to write about certain topics, so each document is a combination of certain topics, and each topic is written using a subset of all possible words. This is called *latent semantic indexing*. It is clear that this makes more sense than trying to come up with rules such as "People who use the word X also use the word Y."

Thinking of how the data is generated through hidden factors and how we believe they combine to generate the observable data is important, and it can make the estimation process much easier. What we discuss here is an additive model where we take a sum of the effects of the hidden factors. Models are not always linear—for example, a factor may inhibit another factor—and learning nonlinear generative models from data is one of the important current research directions in machine learning.

Thinking of how the data is generated through hidden factors and how they combine to generate the observable data is important, and it can make the estimation process much easier.

In the next chapter, we will discuss a different type of scenario where the learning system is an agent that is situated in an environment. The agent, e.g., a robot, has sensors to detect its state in the environment and can take actions, as a result of which it gets a reward or not. As we will see shortly, the aim in this case corresponds to learning what actions should the agent take in which state to maximize the total reward.

LEARNING TO TAKE ACTION

Reinforcement Learning

Let us say we want to build a machine that learns to play chess. Assume we have a camera to see the positions of the pieces on the board, ours and our opponent's, and the aim is to decide on our moves so that we win the game.

In this case, learning is difficult because of two reasons. First, it is very costly to have a teacher who will take us through many games, indicating the best move for each board state. Second, in many cases, there is no such thing as the best move; how good a move is depends on the moves that follow. A single move does not count; a sequence of moves is good if after playing them we win the game. The only real feedback is at the end of the game, it's the result of the game, whether we win or lose.

There is no such thing as the best move. A single move does not count; a sequence of moves is good if after playing them we win the game.

Another example is a robot that is placed in a maze to find a goal location. The robot can move in one of the four compass directions and should make a sequence of movements to reach the goal. There may be obstacles, static or dynamic, that the robot should navigate around. As long as the robot moves around, there is no feedback and the robot tries many moves until it reaches the goal; only then does it get a reward (for correct completion of the task). In this case there is no opponent, but we can have a preference for shorter trajectories—the robot may be running on a battery—which implies that in this case we are playing against time.

These two applications have a number of points in common. There is a decision maker, called the *agent*, which is placed in an *environment* (see figure 17). In the first case,

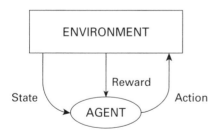

Figure 17   Basic setting for reinforcement learning where the agent interacts with its environment. At any state of the environment, the agent takes an action and the action changes the state and may or may not return a reward.

the chessboard is the environment of the game-playing agent; in the second case, the maze is the environment of the robot. At any time, the environment is in a certain *state*, which means the position of the pieces on the board or the position of the robot in the maze, respectively. The decision maker has a set of *actions* possible: the legal movement of pieces on the chessboard or the movement of the robot in various directions without hitting any obstacle. Once an action is chosen and taken, the state changes.

The solution to the task requires a sequence of actions, and we get feedback in the form of a *reward*. What makes learning challenging is that the reward comes rarely and generally only after the complete sequence has been carried out—we win or lose the game after a long sequence of moves. The reward defines the aim of the task and is necessary if we want learning. The agent learns the best sequence of actions to solve the task where "best" is quantified as the sequence of actions that returns the maximum reward as early as possible. This is the setting of *reinforcement learning* (Sutton and Barto 2018).

Reinforcement learning is different in a number of respects from the learning methods we've already discussed. It is called "learning with a critic," as opposed to the learning with a teacher that we have in supervised learning. A critic differs from a teacher in that a critic does not tell us what to do, but only how well we have been doing in the past. The critic never informs in advance! The feedback is scarce and when it comes, it comes late. This leads to the *credit assignment* problem. After taking several actions and getting the reward, we would like to assess the individual actions we did in the past and find the moves that led us to win the reward so that we can record and recall them later on.

Actually, what a reinforcement learning program does is generate an *internal value* for the intermediate states or actions in terms of how good they are at leading us to the goal and getting us the real reward. Once such an internal reward mechanism is learned, the agent can just take the local actions to maximize it. The solution to the task requires a *sequence* of actions chosen in this way that cumulatively gets the highest real reward.

Unlike the applications we discussed previously, here there is no external process that provides the training data. It is the agent that actively generates data by trying out actions in the environment and receiving feedback (or not) in the form of a reward. It then uses this feedback to update its knowledge so that in time it learns to do actions that return the highest reward.

K-Armed Bandit

We start with a simple example. The *K-armed bandit* is a hypothetical slot machine with *K* levers. The action is to

A reinforcement learning program learns to generate an *internal value* for the intermediate states or actions in terms of how good they are at leading us to the goal and getting us the real reward.

choose and pull one of the levers; each lever returns a certain amount of money, which can be zero, which is the reward associated with the lever (action). The task is to decide which lever to pull to maximize the reward.

This is a classification problem where we choose one of K. If this were supervised learning, the teacher would tell us the correct class, namely, the lever leading to maximum earning. In this case of reinforcement learning, we can only try the different levers and keep track of the best.

Initially estimated values for all levers are zero. To explore the environment, we can choose one of the levers at random and observe its reward. If that reward is higher than zero, we can just store it as our internal reward estimate of that action. Then, when we need to choose a lever again, we can keep on pulling that lever and receiving positive rewards. But it may be the case that another lever leads to a higher reward, so even after finding a lever with a positive reward we want to try out the other levers; we need to make sure that we have done a thorough enough exploration of the alternatives before we become set in our ways. Once we try out all the levers and know everything there is to know, we can then choose the action with the maximum value.

The setting here assumes that rewards are deterministic, that we always receive the same reward for a lever. In a real slot machine, the reward is a matter of chance, and the same lever may lead to different reward values in different trials. In such a case, we want to maximize our *expected reward*, and our internal reward estimate for the action is the average of all rewards in the same situation. This implies that doing an action once is not enough to learn how good it is; we need to do many trials and collect many observations (rewards) to calculate a good estimate of the average.

The K-armed bandit is a simplified reinforcement learning problem because there is only one state—one slot machine. In the general case, when the agent chooses an action, not only does it receive a reward or not, but its state also changes. This next state of the agent may also be probabilistic because of the hidden factors in the environment, and this may lead to different rewards and next states for the same action.

For example, there is randomness in games of chance that also affect the action and thus the next state: in some games there are dice, or we draw randomly from a deck in card games. In a game like chess, there are no dice or decks of cards, but there is an opponent whose behavior is unpredictable—another source of uncertainty. In a robotic environment, the obstacles may move or there may be other mobile agents that can occlude perception or limit movement. Sensors may be noisy and motors that control the actuators may be far from perfect: A robot may want to go ahead, but because of wear and tear may swerve to the right or left. All these are hidden factors that introduce uncertainty, and as usual, we estimate expected values to average out the effect of uncertainty.

Another reason the K-armed bandit is simplified is because we get a reward after a single action; the reward is not delayed and we immediately see the value of our action. In a game of chess or with a robot whose task is to find the goal location in a room, the reward arrives only at the very end, after many actions during which we receive no reward or any other feedback.

In reinforcement learning, what we want is to be able to predict how good any intermediate action is in taking us to the real reward—this is our *internal reward estimate* for the action. Initially, this reward estimate for all actions is zero because we do not yet know anything. We need data to learn, so we need to do some *exploration* where we try out certain actions and observe whether we get any reward; we then update our internal estimates using this information.

As we explore more, we collect more data, and we learn more about the environment and how good our actions are. When we believe we have reached a level where our reward estimates of actions are good enough, we can start *exploitation*. We do this by taking the actions that generate the highest reward according to our internal reward estimates. In the beginning when we do not know much, we try out actions at random; as we learn more, we gradually move from exploration to exploitation by moving from random choices to those influenced by our internal reward estimates.

Temporal Difference Learning

For any state and action, we want to learn the expected cumulative reward starting from that state with that action. This is an *expected* value because it is an average over all sources of randomness in the rewards and the states to come. The expected cumulative rewards of two consecutive state-action pairs are related through the *Bellman equation*, and we use it to *back up* the rewards from later actions to earlier actions, as follows.

In figure 18, we have a grid world. Let us consider the final move of the robot that leads to the goal; because we reach the goal, we receive a reward of, say, 100 units. Now consider the state and action immediately before that. In that state we do an action, which, though it does not give us an immediate reward (because we will still be one step away from the goal), takes us to the state where with one more action we can get the full reward of 100. This means that that action in that state has a lot of value, but it is still one step away. So, to calculate the value, we discount the real reward, let us say by a factor of 0.9 (because the reward is in the future and the future is never certain), and

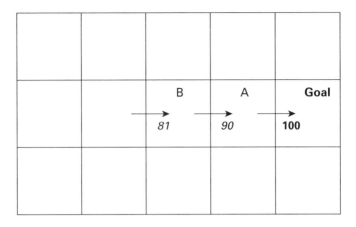

Figure 18 Temporal difference learning through reward backup. When we are in state *A*, if we go right, we get the real reward of 100. In state *B* just before that, if we do the correct action (i.e., go right), we get to *A* where with one more action we can get the real reward, so it is as if going right in *B* also has a reward. But it is discounted (here by a factor of 0.9) because it is one step before, and it is a simulated internal reward, not a real one. The real reward for going from *B* to *A* is zero; the internal reward of 90 indicates how close we are to getting the real reward. Similarly, any action that gets us to *B* has an internal reward of 81.

we say that that particular state-action pair has an *internal reward* of 90.

Note that the real reward there is still zero, because we still have not reached the goal, but we internally reward ourselves for having arrived at a state that is only one step away from the goal. Similarly, the one before that action is discounted twice and gets an internal reward of 81, and we can continue assigning internal values to all the previous

In the beginning, we try out actions at random; as we learn more, we move from exploration to exploitation, from random choices to those influenced by our internal reward estimates.

actions in that sequence. Of course, this is for only one trial episode. We need to do many trials where in each because of the uncertainties we follow a different path visiting different states and observing different rewards, and we average over all those internal reward estimates. This is called *temporal difference* (TD) learning; the internal reward estimate for each state-action pair is denoted by Q, and the algorithm that updates them is called *Q-learning*.

Note that only the final action gets us the real reward; all the values for the intermediate actions are simulated rewards. They are not the aim; they only help us to find the actions that eventually lead us to the real reward. Just like in a school, a student gets grades based on their performance in different courses, but those grades are only simulated rewards indicating how likely it is the student will get the real reward, which they will get only when they graduate and become a productive member of their community.

In certain applications, the environment is *partially observable*, and the agent does not know the state exactly. It is equipped with sensors that return an observation, which it uses to estimate the state of the environment. Let us say we have a robot that navigates in a room. In the preceding grid world example, the robot knows its position exactly, but this may not always be the case. The robot may not know where it is or what else is in the room. The robot may have a camera, but an image does not tell the robot the environment's state in full detail; it only gives some indication about the likely state. For example, the robot may only know that there is an obstacle to its left.

In such a case, based on the observation, the agent predicts its state; or more accurately, it predicts the probability that it is in each state given the observation and then does the update for all probable states weighted by their probabilities. This additional uncertainty makes the task much more difficult and the problem harder to learn (Thrun, Burgard, and Fox 2005).

For example, a self-driving car driving in an urban region knows its environment exactly by accessing its geographic information system data; in a rural region it has to rely more on its onboard sensors for navigation.

Learning to Play Games

One of the early applications of reinforcement learning is the *TD-Gammon* program that learns to play backgammon by playing against itself (Tesauro 1995). This program is superior to the previous *NeuroGammon* program also developed by Tesauro, which was trained in a supervised manner based on plays by experts. Backgammon is a complex task; it features an opponent and extra randomness due to the roll of dice. Using a relatively simple representation of the board, TD-Gammon trains a neural network that is a multilayer perceptron with one hidden layer by playing against a copy of itself.

Previously we talked about calculating the value of an intermediate state by discounting the future reward; a *value network* is a regressor that takes the state as input, here the representation of the backgammon board, and is trained to estimate its value (the expected cumulative reward after that state). A *policy network* that we will see shortly is a classifier that takes the state as input and is trained to choose the best action—namely, the one that takes us to the next state with the maximum value (on the path that returns the maximum expected cumulative reward).

Though reinforcement learning algorithms are slower than supervised learning algorithms, it is clear that they have a wider variety of application and have the potential to construct better learning machines. They do not need any supervision, and this may actually be better since there will not be any teacher bias. For example, Tesauro's TD-Gammon program that learned by playing against itself in certain circumstances came up with moves that turned out to be superior to those made by the best players.

A recent impressive work combines reinforcement learning with deep neural networks to play arcade games (Mnih et al. 2015). The *Deep Q-Network*, which is a policy network, takes directly the 84×84 image of the screen (these are arcade games from the 1980s when image resolution was low) and learns to play the game using only the image and the score information. The network has early convolutional layers for analyzing the image and then fully connected layers to synthesize the best joystick action. Training is end-to-end, from pixels to actions, using a form of Q-learning that we discussed earlier.

What is also interesting is that the same network with the same learning algorithm, network architecture, and hyperparameters can learn any of the 49 games, and on 29 of these it reached or exceeded human performance.

Very recently, the same group developed the AlphaGo system (Silver et al. 2016) that again combines deep convolutional networks with reinforcement learning, this time to play the game of Go. Go is much more difficult than backgammon or chess because the board is larger and there are more moves possible per position, which implies a much larger search space; Go was long believed to be beyond our current computing capabilities.

In AlphaGo, the input is a set of specialized features that represent the 19×19 Go board, and the board is processed by convolutional layers as if it is an image. There is the policy network trained to select the best move and the value network trained to evaluate how close each state is to winning the game. The policy network is first trained with a very large database of expert games and then further improved through reinforcement learning by playing against itself. As we discussed in the preface, AlphaGo defeated the European Go champion, 5 games to 0, in 2015 and defeated one of the greatest Go players in the world, 4 games to 1, in March 2016.

What makes AlphaGo impressive is also the high quality of engineering that went into its implementation, for example, in the way computation is distributed over parallel processing units. AlphaGo has played, and learned from, many more games than any human player can play in a lifetime, and because it learns by playing against a copy of itself, it is playing against a better and better opponent, all the time forcing it to devise cleverer and cleverer strategies to win.

A recent version named AlphaGo Zero (Silver et al. 2017) is trained with less human help. The input is just the raw board without any specialized input features, and there is no initial supervised training with games of expert human players. Another difference is that the policy and value networks are merged into one but deeper network. AlphaGo Zero defeats AlphaGo and is now considered to be the best Go player, human or machine. Recently, the approach was generalized into a single AlphaZero algorithm that can learn to play not only Go but also chess and shogi (Japanese chess) (Silver et al. 2018).

What was helpful in these approaches is that both the arcade game screen and the Go board have a two-dimensional structure that can be analyzed by a convolutional neural network for local features at different levels.

The success of DQN and AlphaGo lies in the way that type of local feature extraction is seamlessly coupled with temporal difference learning using a deep neural network trained end to end.

Reinforcement learning is also applied to card games. *DeepStack* learns to play a two-player variant of poker, named heads-up no-limit Texas hold'em. Unlike backgammon or Go where both parties have full information about the environment (i.e., the board), in poker, in addition to randomness due to draws from a deck, the cards of the opponent are hidden. In this imperfect-information setting, a player needs to make inferences about the opponent's state from their previously observed actions and act accordingly, which allow complicated strategies such as bluffing. DeepStack uses a deep neural network and in a study involving 44,000 hands of poker, defeated professional poker players (Moravcik et al. 2017).

Pluribus learned how to play the six-player variant (where there is not one but five opponents) by playing against five copies of itself, and when playing against five professional human players or with five copies of Pluribus playing against one human, it performed significantly better than humans over the course of 10,000 hands of poker (Brown and Sandholm 2019).

Another significant milestone is *AlphaStar* that learns to play StarCraft, which is a real-time strategy game that involve thousands of decisions and imperfect information.

AlphaStar was rated at Grandmaster level and was above 99.8 percent of officially ranked human players (Vinyals et al. 2019).

Reinforcement Learning in Real Life

An important question nowadays is how we can move from games and use deep reinforcement learning in real-world applications. Games are simplified simulations of real life: its rules for playing, winning, and losing are well defined; when there is randomness (e.g., dice in backgammon), it should be fair to both parties. What also makes games a good testbed for learning is that it is possible to simulate games very fast on a computer and hence collect large amounts of data very quickly.

Real life, in contrast, has all sorts of ambiguities with different sources of uncertainties and sensor noise; actions take time and may be imperfect; losses incurred after bad actions imply monetary costs and may even endanger human safety. In simulating a game, you can try any random action to see what it gets you (and actually the proofs of convergence of the temporal difference algorithms require this), but you cannot do this in real life.

Reinforcement learning is ideally suited to sequential decision-making tasks where we need to generate a sequence of decisions and where each decision affects later decisions. A decision by itself is not good or bad, so we cannot always use supervised learning. But the goodness of a decision depends on all the decisions before and after, so it's the whole sequence of decisions that's being evaluated at the end. There may also be multiple decision-making agents whose actions influence each other's behavior.

There are many scenarios that fit this description (Li 2019). In recommender systems, we need to generate a set of recommendations to each customer; in healthcare, we need to generate the correct sequence of treatment decisions; in economics and finance, we need to generate a good sequence of buy/sell decisions, and so on.

One interesting work is the *neural architecture search* algorithm where designing the best neural network structure is converted to a sequence of decisions on hyperparameters that define the structure and the connectivity, and the reward is the accuracy of this constructed network. So there is the controller neural network that is trained with reinforcement learning, and it learns how to construct the child network one hyperparameter at a time (Zoph and Le 2016).

Another interesting application is regulating the temperature and airflow inside a large-scale data center. The air temperature is regulated through air-water heat exchange, and the controls that can be manipulated are the fan speed (controlling air flow) and the valve opening (for letting cold water in and expelling warm water). It has

Reinforcement learning is ideally suited to sequential decision-making tasks where we need to generate a sequence of decisions and each decision affects later decisions.

been demonstrated that a simple model with little prior knowledge trained using reinforcement learning by a few hours of exploration suffices for the task and is at the same time cost effective (Lazic et al. 2018).

Robotics is an area where reinforcement learning is appropriate because the completion of many robotics tasks require the generation of a sequence of correct actions. Previously we discussed a robot looking for the goal location in a maze, which is a particular case of navigation, where a robot, for example, a self-driving car, needs to find the path from point A to point B subject to constraints such as obstacles, while optimizing a criterion such as time. A multi-legged robot needs to move its legs in the correct order so that it can advance without falling.

One interesting area where robotics meet machine learning is *imitation learning*. Learning a task, if there are many possible actions and if long sequences are needed, requires a lot of exploration and hence can be slow. A blind exploration can also be dangerous because some actions can be harmful to the robot or its environment, so one possibility is to train the robot in a simulated environment, at least in early stages. Or, if we have a person who already knows how to do the task, the robot can learn by imitation. For example, we can have a robot arm learning to manipulate objects in its environment (e.g., learning to put one object on top of another) by watching and imitating a person.

In *behavioral cloning*, the actions taken by the human are recorded step-by-step and taught to the robot; this corresponds to transforming the whole task into a sequence of supervised learning tasks where for each intermediate state, the human action defines the required output. For example, a large data set can be collected by having expert human drivers drive under a wide variety of road, traffic, and weather conditions while recording what actions they take in which situation. Such a data set can then be used to train a driving program.

Another interesting approach for imitation is called *inverse reinforcement learning*, where we first learn a good reward function by observing the human behavior and this reward function is then used to train a standard reinforcement learning program. This is one hot research topic in reinforcement learning. Humans can do a lot of tasks, but they do most tasks without explicitly being aware of how they do it, so imitation learning helps translate this knowledge to a robot.

Machine learning has become an important component of many products and services in the past decade, as a result of which we have also started to notice the associated challenges and risks. We have concerns, for example, about the privacy and security of the data, as well as the transparency, fairness, and accountability of making automated decisions. We cover such concerns in the next chapter.

CHALLENGES AND RISKS

The Other Side of Machine Learning

As any new technology, the use of machine learning introduces new unknowns and possible side effects that need to be spotted and handled appropriately. On the one hand, the success of a machine learning solution depends directly on how much data there is, so we want to have as many users and as much data as possible. But to be able to have that, people should be convinced that the system, be it a product or service, is "on their side"; that it makes lawful and responsible decisions; that the users' right to privacy are respected; that the system is fair to all users and is transparent about how its decisions are made; and that its manufacturers can be held legally accountable when it makes a decision that causes harm. If these conditions are not met, people will be unwilling to provide information

The system, be it a product or service, should make lawful and responsible decisions, respect privacy, be fair to all users and transparent in its operation, and hold its manufacturers accountable.

and as a result there can be no learning and no product or service to sell.

As a general rule, laws are technology neutral. It's a crime to kill a person whether one uses a rock or a drone. The High Court of England and Wales stated in 2019 that "the fact that a technology is new does not mean that it is outside the scope of existing regulation, or that it is always necessary to create a bespoke legal framework for it."[1]

For example, it is a basic human right that there can be no discrimination because of an individual's race, gender, or age. This should be true regardless of who or what is making the decision; so it should be a given that a machine learning system is not allowed to make use of such attributes in making a decision.

This may not be always straightforward when we are learning from large amounts of data, because there can always be correlations that are not immediately apparent. For example, the zip code of an individual can be correlated with and hence can act as a proxy of ethnicity (Favaretto, De Clercq, and Elger 2019). Learning algorithms are surprisingly good at finding such correlations; sometimes, they also find correlations that are spurious, especially when the data set is small. This is an indicator of the need for explainability, which is one requirement that we will discuss shortly.

Automated decision making of course is not new but was used in more limited scenarios in the past; today its application areas are increasing, which is due not only to machine learning but also to the increased precision and decreasing cost of automation.

Airplanes have had autopilots, trains and boats their guidance systems, but building self-driving cars is a much more difficult task; consider an urban environment, which has not only many other cars but also other dynamic agents such as pedestrians, cyclists, and so on. Driving is a mode of transport that is everyday and ubiquitous, so automating it will have more impact. Another example is automated trading where computer programs make buy-or-sell decisions. Face recognition is good if it helps the police to catch criminals; it is bad if it is used to track people without their knowledge and consent. The more widespread use of automation technologies, which depend increasingly on machine learning, is helpful, but comes with downsides, one being that we become reliant on them too much and too quickly.

When a human driver causes an accident, its effect is limited because the probability that another driver makes the exact same erroneous decision is very small; if a self-driving car causes an accident, this means that all other cars of that make are sure to cause an accident if they are put in the same scenario. That is why when an airplane is involved in an accident, all other airplanes of the same make are immediately grounded until the cause of the accident is determined and eliminated; in the future, it may happen that your self-driving car will refuse to run in the morning because some other car has recently been involved in an accident in another city and the reason has not yet been determined.

In addition to risks associated with automated decision making, another source of risk in machine learning is related to data, and this has different aspects. First is the need for data privacy and security. Data may contain personal information that needs to be kept confidential; so any collection, storage, and processing of data should be done with privacy and security concerns in mind. Second is that the quality of any machine learning solution directly depends on the quality of the data. If the data is biased or outright corrupt so will all the decisions based on that data be. Third is the element of trust. In high-risk applications, for example in the medical domain, we require our trained models to be interpretable so that they provide not only a decision but also a human-readable explanation that justifies that particular decision.

Let us now discuss these aspects one by one.

Data Privacy and Security

When we have a lot of data, its analysis can lead to valuable results, and historically, data collection and analysis have resulted in significant findings for humanity, in many domains from medicine to astronomy. The current widespread use of digital technology allows us to collect and analyze the data quickly and accurately, and in numerous new domains.

With more and detailed data, the critical point today is *data privacy and security* (Horvitz and Mulligan 2015). How can we make sure that we collect and process data without infringing on people's privacy concerns and that the data is not used for purposes beyond its original intention?

We expect individuals in a society to be aware of the advantages of data collection and analysis in domains such as health care and safety. And even in other domains such as retail, people always appreciate services and products tailored to their likes and preferences. Still, no one likes to feel that their private life is being pried into. Our smart devices, for example, should not turn into digital paparazzi recording the details of our lives and making them available without our knowledge.

In the past, information about people used to be distributed among data collectors where each had access only to the part that it needed: the bank had data only related to our financial situation, the employer had data only related to our work, and so forth. There was also data related to services provided by different companies, such as a travel agent, or a utility provider, again each seeing a small portion of the person's profile, and our social interactions were not recorded. In a world where these different pieces of data are made easily accessible, even if each one is partially informative and some seemingly unimportant, they can be combined to make very detailed inferences about a person. This is both the power and the risk of learning from data.

The basic requirements in data privacy are that the user who generates the data should always know what and how much data is collected, what part of it is stored, whether that data will be analyzed for any purpose, and if so, what that purpose is. No more data than what is absolutely necessary should be collected. The data collector should be completely open about what is collected.

This requirement of *transparency* implies that the owner of the data should always be informed during both data collection and use. Before any analysis, the data should be sanitized—that is, all the personal details should be hidden to make the record anonymous, which is not a straightforward process. With human records, for instance, just removing unique identifiers such as the name or social security number is not enough; fields such as birth date, zip code, and so on provide partial clues, and individuals can be identified by combining such clues (Sweeney 2002).

Data is becoming such a valuable raw resource that it behooves the collector of the data to take all the necessary steps for its safekeeping and to not share it with someone else without the explicit consent of the data owner.

Individuals should have complete control over their data. They should always have the means to view what data of theirs has been collected; they should be able to ask for its correction or complete removal.

A recent line of research is in *privacy-preserving learning* algorithms. Let us say that we have parts of data from different sources (e.g., different countries may have patients suffering from the same disease) and that they do not want to lend their data (detailed information about their citizens) to a central user to train a model with all the data combined. In such a case, the easiest possibility is to share the data in a form that is sufficiently anonymized, which may not always be easy.

One research topic in machine learning is called *differential privacy*, where the idea is to knowingly corrupt the data before sharing it so that individual records cannot be identified while still allowing correct generalizations to be learned from the whole.

Another approach is called *homomorphic encryption*, where the data is stored encrypted and is also processed without decrypting it. Because the data always remains encrypted, the privacy of the information is preserved; the disadvantage is that this type of processing requires a lot of extra computation.

Yet another interesting idea for privacy is *federated learning*, where the idea is to have multiple copies of the model for different users. Each model then learns from its user and is updated by its local data, also regularly sharing the model updates with the other users. Because the other users see only the model updates and not the data, the data of each user is kept private.

Concerns over data privacy and security should be an integral part of any data analysis scenario, and it should be resolved before any learning is done. Mining data is just like mining for gold—before you start any digging, you need to make sure that you have all the necessary permits. In the future, we may have data processing standards where every data set contains some metadata about this type of ownership and permission information; then, it may be required that any machine learning or data analysis software check for these and run only if the necessary stamps of approval are there.

Biased Data

How good a learned model will be depends first of all on how good its training data is. Any problems with that data will be reflected in the quality of the model.

One source of problems is possible mishaps in collecting the data. For example, if a face recognition data set contains more White faces than those of other races, any face recognizer trained on that data will perform poorly on races that are underrepresented. Basically, the error for any class depends on how many training instances there are of that class, and so the learning algorithm will have less motive to correct the errors on classes that have fewer training instances. This type of sampling error can be corrected by collecting a balanced training sample that includes the whole range of racial identities.

Basically, we need to make sure to match the distribution of the training data with the distribution of the data encountered in the field; any mismatch will result in a poor model. For example, we frequently see that scientific articles that report very high accuracies in the lab do not always lead to successful commercial products. This is partly because the lab conditions are close to ideal, where, for example, data is collected carefully and with high precision; such conditions are almost impossible to duplicate in the real world.

Such erroneous sampling may also have indirect causes such as biases in the underlying process that generates the data. For example, data may have been collected in the past where there may have been lack of diversity; a bank may have had fewer women than men as customers in the past, or fewer customers from certain minorities. Achieving *fairness*, by detecting and mitigating such biases is an important research topic in machine learning.

The fact that a model that learns the general behavior in the data does not make good decisions for underrepresented cases or outliers has other implications as well.

For example, there is an important risk in basing recommendations too much on past use and preferences. If a person only listens to songs similar to the ones they listened to and enjoyed before, or watches movies similar to those they watched and enjoyed before, or reads books similar to the books they read and enjoyed before, then there will be no new experience and that will be limiting, both for the person and for the company that is always eager to find new products to sell. So in any such recommendation scheme, there should also be some attempt at introducing some diversity.

A recent study (Bakshy, Messing, and Adamic 2015) has shown that a similar risk also exists for interactions on social media. If a person follows only those people they agree with and reads posts, messages, and news similar to the ones they have read in the past, they will be unaware of other people's opinions and that will limit their experience, as opposed to traditional news media outlets, such as newspapers or TV, that contain a relatively wider range of news and opinions.

Model Interpretability

As we have more and more computer systems that are trained from data to make autonomous decisions, we need to be concerned with relying so much on computers. One important requirement is the validation and verification of software systems—that is, making sure that they do what they should do and do not do what they should not do.

This may be especially difficult for models trained from data, because training involves different sources of randomness in data and optimization; this makes trained software less predictable than programmed software.

Our approach in testing a trained model is to check its performance on data unused in training, to see how well it has generalized from the particular training examples. After having trained the model on the training data, which is one small random subset of all possible cases, we test it on a validation data that is another random subset, and we use the accuracy on this new data as our criterion of the expected accuracy in later use. This is a useful criterion, but it is not sufficient, because both training and validation sets are small and randomly chosen, and changing them may lead to changes in the accuracy estimates.

A second problem is sensitivity. For example, it has been shown with deep neural networks that slightly perturbed versions of valid examples sometimes cause very big changes at the output. Such *adversarial examples* are taken as an indicator that the model has not generalized well but is still greatly dependent on the individual training instances. This is disturbing because we know that our sensors are never perfect and there can always be noise at the input. For example, we do not want a self-driving car to run amok if bad weather clouds its sensors or the road signs get muddy.

Because of all these reasons, we should always question the decision made by a trained model and require additional validation, especially in domains where wrong decisions have possible high losses. We want our model to provide not only an output but also an explanation as to how it came to that decision. Such explanations should be in a format that does not require any expertise in machine learning, but any user of the system should be able to understand and assess them.

The internal operation of a deep neural network is not easily understandable; it's an example of a *black box* model. However, a linear model where we calculate a total score as a weighted sum of different factors, or a decision tree that can be written down in terms of if-then rules, are easy to understand and hence preferrable (see figure 6). Such rules can be checked and assessed by human experts to see if the trained model has learned meaningful rules. This is the topic of *explainable artificial intelligence* (XAI), which is another research area that is increasingly becoming important as trained models are being used more and more (Gilpin et al. 2018).

One interesting idea is that of "counterfactuals" where the idea is to inform the user how the decision could have been different: for example, "You were denied

We want our model
to provide not only an
output but also an
explanation as to how
it came to that decision,
in a format under-
standable by anyone.

a loan because your annual income was £30,000. If your income had been £45,000, you would have been offered a loan" (Wachter, Mittelstadt, and Russell 2018, 844). Of course, in generating explanations as well as counterfactuals, it is necessary to preserve commercial secrets; banks for example would not want to disclose the exact formula they use for credit scoring.

Ethical, Legal, and Other Social Aspects

As any engineering product, the correct behavior of a trained model should be checked rigorously. We have already considered possible biases in the training data that may be harmful. In any scenario where automatic decisions are made based on past data, such as finance, health care, or justice, we have to guarantee *transparency* in the collection of data and *fairness* in making decisions. We also have to make sure that such systems use models that are interpretable, and that their decisions be *explainable* to its users, and we have to be able to hold the manufacturers of such systems *accountable*.

When intelligence is embodied and the system takes physical actions, the correctness of the behavior becomes an even more critical issue and even human life may be at stake. So another requirement is *safety*. The system does not need to be a drone with onboard weaponry for this to be true; even an autonomous car becomes dangerous if it is driven badly. When such concerns come into play, the usual expected value or utilitarian approaches do not apply, as is discussed in the "trolley problems," a variant of which is as follows.

Let us say you are riding in an autonomous car when a child suddenly runs across the road. Assume the car is going so fast that it knows it cannot stop. But it can still steer, and it can steer to the right to avoid hitting the child. But let us say that the child's mother is standing to the right of the road. How should the car decide? Should it go ahead and hit the child, or steer right and hit the mom instead? How can we program such a decision? Or should the car instead steer left and drive off the cliff after calculating that your life is worth less than that of the child or the mother's? Can the driving software be allowed to take factors such as age or gender into account in making a decision?

This seems like an extreme case—trolley problems are thought experiments—however, the point is that in many decision-making scenarios, there are multiple possible actions with possibly harmful results, and we need to come up with a way to program such decisions in machines that is in line with our customary standards of ethics and morality.

Increased intelligence due to machine learning can also be used for outright malicious and criminal practices.

Pattern recognition technologies such as face or speech recognition can be used for mass surveillance. Intelligent robots can use their intelligence to become better at killing. Collected data can be analyzed for subversive purposes, such as to influence voter decisions. These are some other ethical and legal implications of the use of machine learning.

Another aspect is that making machines more intelligent leads to higher automation and therefore job loss. Before, jobs became outdated from one generation to the next—for example, sons of blacksmiths became car mechanics—but now it's happening within a generation. Once, people learned a job in their twenties that they did with small updates until they retired. Now it seems as if we need to learn a new job every ten or fifteen years. To be able to catch up with increasing intelligence of the machines, we need to continuously increase our knowledge as well, which means that lifelong education programs are going to become more important in the future. This also implies that unemployment should lose its social stigma but be treated as a normal state of affairs; that time in between is the period when the person is learning their next job, their next version, so to speak.

The next, final chapter will discuss possible future directions of machine learning research and applications.

WHERE DO WE GO FROM HERE?

Make Them Smart, Make Them Learn

Machine learning has already proved itself to be a viable technology, and its applications in many domains are increasing every day. This trend of collecting and learning from data is expected to continue even stronger in the near future (Jordan and Mitchell 2015). Analysis of data allows us to both understand the process that underlies the past data—just like scientists have been doing in different domains of science for hundreds of years—and also predict the behavior of the process in the future.

A few decades ago, computing hardware used to advance one microprocessor at a time; every new microprocessor—first size 8, then 16, then 32 bits—could do slightly more computation in unit time and could use slightly more memory, and this translated to slightly better computers. Similarly, computer software used to advance one programming language at a time. Each new language made some new type of computation easier to program. When computers were used for number crunching, we programmed in Fortran; when we used them for business applications, we used Cobol; later on, when computers started to process all types of and more complex types of information, we developed object-oriented languages that allowed us to define more complicated data structures together with the specialized algorithms that manipulate them.

Then, computing started to advance one operating system at a time; each new version made computers easier to use and supported a new set of applications. Today, computing is advancing one smart device or one smart app at a time. Once, the key person who defined the advance of computing was the hardware designer, then it was the software engineer, then it was the user sitting in front of their computer, and now it is anyone while doing anything.

Nobody eagerly awaits a new microprocessor anymore, and neither a new programming language nor a new operating system version is newsworthy. Now we wait for the next new device or app, smart either because of its designer, or because it learns to be smart.

More and more of our lives are being projected in the digital domain, and as a result we are creating more and more data. The earliest hard disks for personal computers had a capacity of five megabytes; now a typical computer comes with five hundred gigabytes—this is one hundred thousand times more storage capacity in roughly thirty years. A reasonably large database now stores hundreds of terabytes, and we have already started using the petabyte as a measure; very soon, we will jump to the next measure, the exabyte, which is one thousand petabytes or one million terabytes. Together with an increase in storage capacity, processing has also become cheaper and faster thanks to advances in technology that deliver both faster computer chips and parallel architectures containing thousands of processors that run simultaneously, each solving one part of a large problem.

The trend of moving from all-purpose personal computers to specialized smart devices is also expected to accelerate. We discussed in chapter 1 how in the past organizations moved from a computer center to a distributed scheme with many interconnected computers and storage devices; now a similar transformation is taking place for a single user. A person no longer has one personal computer that holds all their data and does all their processing; instead, their data is stored in the "cloud," in some remote offsite data center, but in such a way as to be accessible from all their smart devices, each of which accesses the part it needs.

What we call the *cloud* is a virtual computer center that handles all our computing needs. We do not need to worry about where and how the processing is done or where and how the data is stored as long as we can keep accessing it whenever we want. This used to be called *grid computing*, analogous to the electrical grid made up of an interconnected set of power generators and consumers; as consumers, we plug our TV in the nearest outlet without a second thought as to where the electricity comes from.

This also implies connectivity with larger bandwidths. Streaming music and video is already a feasible technology today. CDs and DVDs we keep on our shelves (that had once supplanted the nondigital LPs and videotapes) have now in turn become useless and are replaced by some invisible source that stores *all* the songs and the movies. E-book and digital subscription services are quickly replacing the printed book and the bookstore, and search engines have long ago made window stoppers out of thick encyclopedias.

With smart devices, there is no longer any need for millions of people to store separate copies of the same song/movie/book locally. The motto now is: Do not buy it, rent it! Buy the smart device, or the app, or the subscription to the service, and the bandwidth that allows you to access it when you need it.

This change and ease in accessibility also offers new ways to "package" and sell products. For example, traditionally music LPs and CDs corresponded to an "album" that is made up of a number of songs; it is now possible to rent individual songs. Similarly, it is now possible to purchase a single short story without buying the book of collected stories.

In chapter 5, we discussed the use of machine learning in recommendation systems. With more shared data and streaming, there will be more data to analyze, and furthermore, the data will be more detailed. For example, now we also know how many times a person listened to a song or how far they watched into a movie, and such information can be used as a measure of how much the person enjoyed the product.

With advances in mobile technology, there is continuing interest in wearable devices. The smartphone, a wearable device, is now much more than a phone; it also acts as an intermediary for smaller smart "things," such as a watch or glasses, by putting them online. The phone may become even smarter in the near future, for example, with an app for real-time translation: You'll speak in your own language on one end and the person on the other end will hear it automatically translated into their own tongue, not only with the content syntactically and semantically correct but also in your voice and uttered with correct emphasis and intonation.

Machine learning will help us make sense of an increasingly complex world. Already we are exposed to more data than what our sensors can cope with or our brains can process. Information repositories available online today contain massive amounts of digital text and are now so big

Machine learning will help us make sense of an increasingly complex world. Already we are exposed to more data than what our sensors can cope with or our brains can process.

that they cannot be processed manually. Using machine learning for this purpose is called *machine reading*.

We need search engines that are smarter than the ones that use just keywords. Currently we have information distributed in different sources or mediums, so we need to query them all and merge the responses in an intelligent manner. These different sources may be in different languages—for example, a French source may contain more information on the topic even though your query is in English. A query may also trigger a search in an image or video database. And still, the overall result should be summarized and condensed enough to be digestible by a user.

Web scraping is when programs automatically surf the web and extract information from web pages. These web pages may be social media, and accumulated information can be analyzed by learning algorithms, for instance, to track trending topics and the detection of sentiment, opinions, and beliefs about products and people—for example, politicians in election times. Another important research area where machine learning is used in social media is to identify the "social networks" of people who are connected; analyzing such networks allows us to find cliques of like-minded individuals, or to track how information propagates over social media.

One of the current research directions is in adding smartness—that is, the ability to collect, process, and make inferences from data, as well as to share it with other online devices—to all sorts of traditional tools and devices, including traditional wearables such as glasses and watches. When more devices are smart, there will be more data to analyze and make meaningful inferences from. Different devices and sensors collect different aspects of the task, and a critical task will be to combine and integrate these multiple modalities. This implies all sorts of new interesting scenarios and applications where learning algorithms can be used.

Smart devices can help us both at work and at home. Machine learning helps us in building systems that can learn their environment and adapt to their users, to be able to work with minimum supervision and maximum user satisfaction.

Important work is being done in the field of smart cars. Cars that are online allow their passengers to be online and can deliver all types of online services, such as streaming video, over their digital *infotainment* systems. Cars that are online can also exchange data for maintenance purposes and access real-time information about the road and weather conditions. If you are driving under difficult conditions, a car that is a mile ahead of you is a sensor that is a mile ahead of you.

But more important than being online is when cars will be smart enough to help with the driving itself. Cars already had assistance systems for cruise control, self-parking, and lane keeping, but these days they are becoming even more capable. The ultimate aim is for them to completely take over the task of driving, and to that end we already have prototypes of such autonomous vehicles today. Of course, this is also true also for buses, trucks, and so on. For example, in the case of a pandemic when the population is on lockdown, self-driving trucks can transport products from factories to cities.

The visual system of a human driver does not have a very high resolution, and they can only look in a forward direction. Though their visual field is slightly extended through the use of side and rearview mirrors, blind spots remain. A *self-driving car*, on the other hand, can have cameras with higher resolution in all directions and can also use sensors that a human does not have, such as GPS, ultrasound, or night vision, or it can be equipped with a special type of radar, called LIDAR, that uses a laser for measuring distance. A smart car can also access all sorts of extra information, such as the weather, much faster. An electronic driver has a much shorter reaction time.

Machine learning plays a significant role in self-driving cars that result in both smoother driving, faster control, and greater fuel efficiency, but also in smart sensing, for example, by automatic recognition of pedestrians, cyclists, traffic signs, and so forth. Self-driving cars will not only be safer but they will also be faster; the speed limits we have are set because of the relatively slower reaction times of human drivers. There are still problems though:

Lasers and cameras are not very effective in harsh weather conditions—when there is rain, fog, or snow—so technology should advance until we get smart cars that can run in all types of weather.

Self-driving cars and robot taxis are expected to take over driving in cities and on highways in the next decade. It also seems very likely that sometime in the next decade or so, cars and drones will fuse and we will have self-piloting flying cars, with their concomitant tasks that will be best handled by machine learning.

Machine learning has the basic advantage that a task does not need to be explicitly programmed but can be learned. Space will be the new frontier for machine learning as well. Future space missions will very likely be unmanned. Before, we needed to send humans because we did not have machines that were as smart and versatile, but now we have capable robots. If there are no humans on board, the load will be lighter and simpler, and there will be no need to bring the load back. If a robot is to boldly go where no one has gone before, it can only be a learning robot.

High-Performance Computation

With big and bigger data, we need storage systems that have higher capacity *and* faster access. Processing power will necessarily increase so that more data can be processed in a reasonable time. This implies the need for high-performance computer systems that can store a lot of data and do a lot of computation very quickly.

There are physical limits such as the speed of light and the size of the atom, which suggests an upper limit on the speed of transfer[1] and a lower limit on the size of the basic electronics. The obvious solution to this is parallel processing—if you have eight lines in parallel, you can send eight data items at the same time; and if you have eight processors, you can process those eight items simultaneously, in the time it takes to process a single one.

Today parallel processing is routinely used in computer systems. We have powerful computers that contain thousands of processors running simultaneously. There are also multicore machines where a single computing element has multiple "cores" that can do simple computations simultaneously, implementing parallel processing in a single physical chip.

But high-performance computation is not just a hardware problem; we also need good software interfaces to distribute the computation and data over a very large number of processors and storage devices. Indeed, software and hardware for parallel and distributed computation for big data are important research areas in computer science and engineering today.

In machine learning, the parallelization of learning algorithms is becoming increasingly important. Models can be trained in parallel over different parts of the data on different computers and then these models can be merged. Another possibility is to distribute the processing of a single model over multiple processors. For example, with a deep neural network composed of thousands of units in multiple layers, different processors can execute different layers or subsets of layers working in a pipeline manner.

The *graphical processing unit* (GPU) was originally made for rapid processing and the transfer of images in graphical interfaces—for example, in video game consoles—but the type of parallel computation and transfer used for graphics has also made them suited for many machine learning tasks. Indeed, specialized software libraries are being developed for this purpose and GPUs are frequently used by researchers and practitioners effectively in various machine learning applications; for example, the AlphaGo network that we discussed in chapter 6 is parallelized to run on GPUs. There is also research in developing more specialized processing units, for example, to carry out the sort of calculations used in neural networks; today's deep neural networks with hundreds of layers and millions of parameters run too slowly on an ordinary CPU.

We are seeing a trend toward *cloud computing* in machine learning applications too, where instead of buying and maintaining the necessary hardware, people rent the use of offsite *data centers*. A data center is a physical site that houses a very large number of computing servers with many processors and ample storage. There are typically multiple data centers in physically different locations; they are all connected over a network, and the tasks are automatically distributed and migrated from one to the other, so that the load from different customers at different times and in different sizes is balanced. All of these requirements fuel significant research today.

One important use of the cloud is in extending the capability of smart devices, especially the mobile ones. These online, low-capacity devices can access the cloud from anywhere to exchange data or request computation that is too large or complex to do locally. Consider speech recognition on a smartphone. The phone captures the acoustic data, extracts the basic features, and sends them to the cloud. The actual recognition is done in the cloud and the result is sent back to the phone.

In computing, there are two parallel trends. One is in building general-purpose computers that can be programmed for different tasks and for different purposes, such as those used in servers in data centers. The other is to build specialized computing devices for particular tasks, packaged together with specialized input and output. The latter used to be called *embedded systems* but are today called *cyber-physical systems*, to emphasize the fact that they work in the physical world with which they interact.

A system may be composed of multiple such units (some of which may be mobile), and they may be interconnected over a network—for example, a car, a plane, or a home may contain a multitude of such devices for different tasks. Making such systems smart—in other words, able to adapt to their particular environment, which includes the user—is an important research direction.

Following this idea, one popular research topic these days is *edge computing*, where we want to have as much of the specialized processing as possible done "on the edge"—that is, closer to where the data originates. With computation getting cheaper and smaller, most of the necessary computation can be done locally as soon as it is sensed, which has the advantages that we do not need to transmit data back and forth, so there is less network traffic and hence the response is also faster. This is especially interesting in artificial intelligence where we have large chunks of data such as video, image, or sound, and processing them on the spot pushes intelligence to the edge; hence the name "edge AI."

A related concept is *fog computing*, where we have general computing services, just like in cloud computing, but they are closer to the user. For example, they may be in our local area network as opposed to in a faraway data center, similar to how fog is a thin cloud closer to the earth, or us. Again, the advantages are less communication and faster decision making.

How Green Is My AI?

Recently with the proliferation of personal computers, smart phones, Internet, and data centers, the total amount of electricity used to power the computers around the globe has reached a considerable amount. Computing used to be considered environmentally friendly—it is better to read from the screen than to print it on paper—which is partially true, but we always need to keep in mind that all our calculations, data storage, and communication run on electricity. All those machines need to be powered up and they need to be cooled down, which implies an ever-increasing carbon footprint.

Machine learning is particularly power-hungry. We need to store large data sets and typically learning algorithms need to do a large number of learning iterations, each of which takes a lot of computation, and hence power, when we use a complex model such as a deep neural network with many layers. Schwartz et al. (2019) report that "the computations required for deep learning research have been doubling every few months, resulting in an estimated 300,000x increase from 2012 to 2018."

This has a number of disadvantages: First, if a model uses too much power, it cannot be implemented on a mobile device running on a battery. Second, a model that uses too much computation will be expensive and such a device cannot be sold to a large customer base. Third, all that power needs to be produced somehow, and we know that frequently countries need to burn coal or gas to generate that electricity, which has all sorts of detrimental effects on the environment including the contribution to global warming.

Research on more energy-efficient computer architectures is an important topic for computing, and energy efficiency has become an important criterion in assessing the quality of an algorithm; this is especially relevant for machine learning that is both data- and computation-heavy.

Data Mining

Though the most important, machine learning is only one step in a *data mining* application (Han and Kamber 2011). There is also the preparation of data beforehand and the interpretation of the results afterward.

Making data ready for mining involves several stages. First, from a large database with many fields, we select the parts that we are interested in and create a smaller database to work with. It may also be the case that the data comes from different databases, so we need to merge them. The level of detail may also be different—for instance, from an operational database we may extract daily sums and use those rather than the individual transactions. Raw data may contain errors and inconsistencies or parts of it may

Energy efficiency has become an important criterion in assessing the quality of an algorithm; this is especially relevant for machine learning that is both data- and computation-heavy.

be missing, and those should be handled beforehand in a preprocessing stage.

After extraction, data is stored in a *data warehouse* on which we do our analysis. One type of data analysis is manual where we have a hypothesis—"people who buy *X* also buy *Y*"—and check whether the data supports the hypothesis. The data is in the form of a spreadsheet where the rows are the data instances—baskets—and the columns are the attributes—products. One way of conceptualizing the data is in the form of a *multidimensional data cube* whose dimensions are the attributes, and data analysis operations are defined as operations on the cube, such as slice, dice, and so on. Such manual analysis of the data as well as visualization of results is made easy by *online analytical processing* (OLAP) tools.

OLAP is restrictive in the sense that it is human-driven, and we can only test the hypotheses we can imagine. For example, in the context of basket analysis, we cannot find any relationship between distant pairs of products; such discoveries require a data-driven analysis, as is done by machine learning algorithms.

We can use any of the methods we discussed in previous chapters, for classification, regression, clustering, and so on, to build a model from the data. Typically, we divide our data into two as a training set and a validation set. We use the first part for training our model and then we measure its prediction accuracy on the validation set. By testing on instances not used for training, we want to estimate how well the trained model would do if used later on, in the real world. The validation set accuracy is one of our main criteria in accepting or rejecting the trained model.

In the previous chapter, we covered the interpretability of machine learning models, and this is an important requirement in data mining. People who use the predictive models do not always know machine learning, so it is important that whatever is learned from the data be presented in a form that is understandable by them. In many data mining scenarios—for example, in credit scoring—this process of knowledge extraction and model assessment by people may be important and even necessary in validating the model trained from data.

Visualization tools can also help here. Actually, visualization is one of the best tools for data analysis, and sometimes just visualizing the data in a smart way is enough to understand the characteristics of the process that underlies a complicated data set, without any need for further complex and costly statistical processing; see Börner 2015 for examples.

As we have more data and more computing power, we can attempt more complicated data mining tasks that try to discover hidden relationships in more complex scenarios. Most data mining tasks today work in a single domain using a single source of data. Especially interesting is the case where we have data from different sources in different modalities; mining such data and finding dependencies across sources and modalities is a promising research direction.

Data Science

The advances and successes of machine learning methods on big data and the promise of more have prompted researchers and practitioners in the industry to call this endeavor a new branch of science and engineering. There are still discussions about what this new field of *data science* should cover, but it seems as if the major topics are machine learning, high-performance computing, and the social, ethical, and legal implications of data collection, analysis, and data-driven decision making.

Of course, not all learning applications need a cloud, or a data center, or a cluster of computers. One should always be wary of hype and companies' sale strategies to invent new and fancier names under which to sell old products.

However, when there is a lot of data and the process involves a lot of computation, efficient implementation of machine learning solutions is an important matter.[2] Another integral part is the ethical and legal implications of data analysis and processing, as we discussed in chapter 7. As we collect and analyze more and more data, our decisions in various domains will become more and more automated and data-driven, and we need to be concerned about the implications of such autonomous processes and the decisions they make.

It seems as if we will need many "data scientists" and "data engineers" in the future, because we see today that the importance of data and extracting information from data has been noticed in many domains. Such scenarios have characteristics that are drastically different than those of traditional statistics applications.

First, the data now is much bigger—consider all the transactions done at a supermarket chain. For each instance, we have thousands of attributes—consider a gene sequence. The data is not just numbers anymore; it consists of text, image, audio, video, ranks, frequencies, gene sequences, sensor arrays, click logs, lists of recommendations, and so on. Most of the time data does not obey the parametric assumptions, such as the bell-shaped Gaussian curve, that we frequently use to make estimation easier. Instead, with the new data, we need to resort to more flexible nonparametric models whose complexity can adjust automatically to the complexity of the task underlying the data. All these requirements make machine learning more challenging than statistics as we used to know and practice it.

In education, this implies that we need to extend the courses on statistics to cover these additional needs, and teach more than the well-known but now insufficient, mostly univariate (having a single input attribute) parametric methods for estimation, hypothesis testing, and regression. It has also become necessary to teach the basics of high-performance computing, both the hardware and the software aspects, because in real-world applications how efficiently the data is stored and manipulated may be as critical as the prediction accuracy. A student of data science today also needs to know the social, ethical, and legal aspects of all stages of machine learning, including the collection of data, its storage and processing, and automated decision making based on that data.

Machine Learning, Artificial Intelligence, and the Future

Machine learning is one way to achieve artificial intelligence. By training on a data set, or by repeated trials using reinforcement learning, we can have a computer program behaving so as to maximize a performance criterion, which in a certain context appears intelligent.

One important point is that intelligence is a vague term and its applicability to assess the performance of computer systems may be misleading. For example, evaluating computers on tasks that are difficult for humans, such as playing chess, is not a good idea for assessing intelligence. Chess is a difficult task for humans because it requires deliberation and planning, whereas humans, just like other animals, have evolved to make very quick decisions using limited sensory data with limited computation. For a computer, it is much more difficult to recognize the face of its opponent than to play chess. Whether a computer can play chess better than the best human player is not a good indicator that computers are more intelligent, because human intelligence has not evolved for tasks like chess.

Researchers use game playing as a testing area in artificial intelligence because games are relatively easy to define with their formal rules and clearly specified criteria for winning and losing. There are a certain number of pieces or cards, and even if there is randomness its form is well defined: the dice should be fair and draws from the deck should be uniform. Attempts to the contrary are considered cheating behavior. In real life, all sorts of randomness occur, and for its survival every species is slowly evolving to be a better cheater than the rest.

The power that artificial intelligence promises is a concern for many researchers, and not surprisingly there is a call for regulation. In a recent interview (Bohannon 2015), Stuart Russell, a prominent researcher and coauthor of the leading textbook on artificial intelligence (Russell and Norvig 2020), says that unlimited intelligence may be as dangerous as unlimited energy and that uncontrolled artificial intelligence may be as dangerous as nuclear weapons. The challenge is to make sure that this new source of intelligence is used for good and not for bad, to increase the well-being of people and for the benefit of humanity, rather than to increase the profit of a few.

Some people jump to conclusions and fear that research on artificial intelligence may one day lead to metallic monsters that will rise to dominate us—electronic versions of the creation of Dr. Frankenstein. I doubt whether that will ever happen. But even today we have automatic systems that make decisions for us—some of which may be trained from data—in various applications from cars to trading. I believe we have more reason to fear the poorly programmed or poorly trained software than we do to dread the possibility of the dawn of super-intelligent machines.

Closing Remarks

We have big data, but tomorrow's data will be bigger. Our sensors are getting cheaper and hence being used more widely and more precisely. Computers are getting bigger too, in terms of their computing power. We still seem to be far from the limits imposed by physics as researchers find new technologies and materials, such as the graphene, that promise to deliver more. New products can be designed and produced much faster using 3D printing technology and more of these products will need to be smart.

With more data and computation, our trained models can get more and more intelligent. Current deep networks can learn enough abstraction in some limited context to recognize handwritten digits or a subset of objects, but they are far from having the capability of our visual cortex to recognize a scene—one deep network does not a brain make. They can learn some linguistic abstraction from large bodies of text, but we are far from any real understanding of it—enough, for example, to answer questions about a short story. How our learning algorithms will scale up is an open question. That is, can we train a model that is as good as the visual cortex by adding more and more layers to a deep network and training it with more and more data? Can we get a model to translate from one language to another by having a very large model trained with a lot of data? The answer should be yes, because our brains are such models. But this scaling up may be increasingly difficult. Even though we are born with the specialized hardware, it still takes years of observing our environment before we utter our first sentence.

In vision, as we go from barcode to optical character readers to face recognizers, we define a sequence of increasingly complex tasks, each of which solves a need and each of which is a marketable product in its own time. More than scientific curiosity, it is this process of capitalization that fuels research and development. As our learning systems get more intelligent, they will find use in increasingly smarter products and services.

In the last half century, we have seen that as computers find new applications in our lives, they have also changed our lives to make computation easier. Similarly, as our devices get smarter, the environment in which we live, and our lives in it, will change. Each age uses its current technology, which defines an environment with its constraints, and these propel new inventions and new technologies. If we can go back two thousand years and somehow give Romans the cell phone technology, I doubt that it would greatly enhance their quality of life, when they were still riding horses, that is, when the rest of their lives did not match up. The world when we will need human-level intelligence in machines will be a very different world.

When will we reach that level of intelligence and how much processing and training will be required are yet to be seen. Currently machine learning seems to be the most promising way to achieve it, so stay tuned.

Adversarial example
A slightly perturbed example that causes a big change at the output. Adversarial examples are an indicator that the model's response is very much specialized to the training examples, and that the model has not correctly generalized.

Anonymization
Removal or hiding of information such that the source cannot be uniquely identified. It is not as straightforward as one would think.

Artificial intelligence
Programming computers to do things, which, if done by humans, would be said to require "intelligence." It is a human-centric and ambiguous term: calling computers "artificially intelligent" is like calling driving "artificial running."

Association rules
If-then rules associating two or more items in *basket analysis*. For example, "People who buy diapers frequently also buy beer."

Autoencoder network
A type of *neural network* that is trained to reconstruct its input at its output. Because there are fewer intermediary hidden units than inputs, the network is forced to learn a short, compressed representation at the hidden units, which can be interpreted as a process of abstraction.

Backpropagation
A learning algorithm for artificial neural networks used for *supervised learning*, where connection weights are iteratively updated to decrease the approximation error at the output units.

Bag of words
A method for document representation where we preselect a lexicon of N words and we represent each document by a list of length N where element i is 1 if word i exists in the document and is 0 otherwise.

Basket analysis
A basket is a set of items purchased together (e.g., in a supermarket). Basket analysis is finding items frequently occurring in the same basket. Such dependencies between items are represented by *association rules*.

Bayes' rule
One of the pillars of probability theory where for two or more random variables that are not independent, we write conditional probability in one direction in terms of the conditional probability in the other direction:

$P(B|A) = P(A|B)P(B)/P(A)$.

It is used, for example, in diagnosis where we are given $P(A|B)$ and B is the cause of A. Calculating $P(B|A)$ allows a diagnostics—that is, the calculation of the probability of the cause B given the symptoms A.

Bayesian estimation
A method for parameter estimation where we use not only the sample, but also the prior information about the unknown parameters given by a *prior distribution*. This is combined with the information in the data to calculate a *posterior distribution* using Bayes' rule.

Bayesian network
See *graphical model*.

Behavioral cloning
One way of doing *imitation learning* where we observe how a human is solving the task step by step and each step is learned in a supervised manner; the robot learns to copy the human behavior exactly—that is, what the correct action is for each intermediate stage.

Bioinformatics
Computational methods, including those that use machine learning, for analyzing and processing biological data.

Biometrics
Recognition or authentication of people using their physiological characteristics (e.g., face, fingerprint) and behavioral characteristics (e.g., signature, gait).

Character recognition
Recognizing printed or handwritten text. In optical recognition, the input is visual and is sensed by a camera or scanner. In pen-based recognition, the writing is done on a touch-sensitive surface and the input is a temporal sequence of coordinates of the pen tip.

Class
A set of instances having the same identity. For example, 'A' and '*A*' belong to the same class. In machine learning, for each class we learn a *discriminant* from the set of its examples.

Classification
Assignment of a given instance to one of a set of *classes*.

Cloud computing
A recent paradigm in computing where data and computation are not local but handled in some remote off-site data center. Typically there are many such data centers, and the tasks of different users are distributed over them in a way invisible to the user. This was previously called grid computing.

Clustering
Grouping of similar instances into clusters. This is an *unsupervised learning* method because the instances that form a cluster are found based on their similarity to each other, as opposed to a *classification* task where the supervisor assigns instances to classes by explicitly labeling them.

Connectionism
A neural network approach in cognitive science where mind is modeled as the operation of a network of many simple processing units running in parallel. Also known as *parallel distributed processing*.

Cyber-physical systems
Computational elements directly interacting with the physical world. Some may be mobile. They may be organized as a network to handle the task in a collaborative manner. Also known as *embedded systems*.

Data analysis
Computational methods for extracting information from large amounts of data. *Data mining* uses machine learning and is more data-driven; *OLAP* is more user-driven.

Data mining

Machine learning and statistical methods for extracting information from large amounts of data. For example, in *basket analysis*, by analyzing large number of transactions, we find *association rules*.

Data science

A recently proposed field in computer science and engineering composed of machine learning, high performance computing, and social, ethical, and legal aspects of data collection and analysis. Data science is proposed to handle in a systematic way the "big data" problems that face us today in many different scenarios.

Data warehouse

A subset of data selected, extracted, and organized for a specific data analysis task. The original data may be very detailed and may lie in several different operational databases. The warehouse merges and summarizes them. The warehouse is read-only; it is used to get a high-level overview of the process that underlies the data either through *OLAP* and visualization tools, or by *data mining* software.

Database

Software for storing and processing digitally represented information efficiently.

Decision tree

A hierarchical model composed of decision nodes and leaves. The decision tree works fast, and it can be converted to a set of *if-then rules*, and as such allows *knowledge extraction*.

Deep learning

Methods that are used to train models with several levels of abstraction from the raw input to the output. For example, in visual recognition, the lowest level is an image composed of pixels. In layers as we go up, a deep learner combines them to form strokes and edges of different orientations, which can then be combined to detect longer lines, arcs, corners, and junctions, which in turn can be combined to form rectangles, circles, and so on. The units of each layer may be thought of as a set of primitives at a different level of abstraction.

Deep Q-Network
A deep neural network trained end to end with *Q-learning*.

Dimensionality reduction
Methods for decreasing the number of input attributes. In an application, some of the inputs may not be informative, and some may correspond to different ways of giving the same information. Reducing the number of inputs also reduces the complexity of the learned model and makes training easier. See *feature selection* and *feature extraction*.

Discriminant
A rule that defines the conditions for an instance to be an element of a *class* and as such separates them from instances of other classes.

Document categorization
Classification of text documents, generally based on the words that occur in the text (e.g., using *bag of words* representation). For instance, news documents can be classified as politics, arts, sports, and so on; emails can be classified as spam versus not-spam.

Edge computing
Processing data at the "edge," that is, where the data is collected, instead of sending it to the cloud to be processed. It leads to fast response and decreased network traffic. This is an idea similar to *fog computing*.

Embedded systems
See *cyber-physical systems*.

Face recognition
Recognizing people's identities from their face images captured by a camera.

Feature extraction
As a method for *dimensionality reduction*, several original inputs are combined to define new, more informative features.

Feature selection
A method that discards the uninformative features and keeps only those that are informative; it is used for *dimensionality reduction*.

Fog computing
A fog is like a cloud, but it is smaller and local. The cloud can be far; but the machines that make up the fog are local, therefore they lead to faster response; the idea is similar to *edge computing*.

Gating unit
A unit that opens or closes another connection depending on its input. It thus allows selectively turning on/off parts of a *neural network*.

Generalization
How well a model trained on a training set performs on new data unseen during training. This is at the core of machine learning. In an exam, a teacher asks questions that are different from the exercises already solved while teaching the course, and students' performance is measured by their performance on these new questions. A student who can solve only the questions that the instructor has solved in class is not good enough.

Generative model
A model defined in such a way so as to represent the way we believe the data has been generated. We think of hidden causes that generate the data and also of higher-level hidden causes. Slippery roads may cause accidents, and rain may have caused roads to be slippery.

Generative adversarial network
This is actually made up of two networks, a generator G and a discriminator D. G generates a "fake" instance from a random input and D is trained to separate such fakes from true examples. G is in turn trained to generate fakes that D will classify as true, which will force D to get better at spotting fakes, which will in turn cause G to be a better faker, and so on.

Graphical model
A model representing dependencies between probabilistic concepts. Each node is a concept with a different truth degree and a connection between nodes represents a conditional dependency. If I know that the rain causes my grass to get wet, I define one node for rain and one node for wet grass, and I put a directed connection from the rain node to the node for wet grass. Probabilistic inference on such networks may be implemented as efficient graph algorithms. Such a network is a visual representation and helps understanding. Also known as a *Bayesian network*—one rule of inference used in such networks is *Bayes' rule*.

High-performance computing
To handle the big data problems we have today in reasonable time, we need powerful computing systems, both for storage and calculation. The field of high-performance computing includes work along these directions; one approach is *cloud computing*.

If-then rules
Decision rules written in the form of "IF antecedent THEN consequent." The antecedent is a logical condition and if holds true for the input, the action in the consequent is carried out. In *supervised learning*, the consequent corresponds to choosing a certain output. A rule base is composed of many if-then rules. A model that can be written as a set of if-then rules is easy to understand and hence rule bases allow *knowledge extraction*.

Ill-posed problem
A problem where the data is not sufficient to find a unique solution. Fitting a model to data is an ill-posed problem, and we need to introduce *inductive bias* to get a final model.

Imitation learning
In robotics, this means training a robot to imitate a human doing the task.

Induction
Learning a general model from particular examples, for example, learning the general concept of a chair from all the chairs one sees.

Inductive bias
The set of assumptions that each machine learning algorithm makes, in addition to the data, to learn a model.

Information retrieval
Given a database of many records, we make a query and we want the records relevant to be found. For example, given a database of news articles, we can make a query using keywords.

Inverse reinforcement learning
As a method for *imitation learning*, the idea is to first an extract reward function by observing the way a human is solving a task; once such a reward function is extracted, we can use *reinforcement learning* proper.

Knowledge extraction

In some applications, notably in data mining, after training a model, we would like to be able to understand what the model has learned; this can be used for validating the model by people who are experts in that application, and it also helps to understand the process that generated the data. Some models are "black box" in that they are not easy to understand; some models—for example, linear models and decision trees—are interpretable and allow extracting knowledge from a trained model.

Latent semantic analysis

A learning method where the aim is to find a small set of hidden (latent) variables that represent the dependencies in a large sample of observed data. Such hidden variables may correspond to abstract (e.g., semantic) concepts. For example, each news article can be said to include a number of "topics," and although this topic information is not given explicitly in a supervised way in the data, we can learn them from data such that each topic is defined by a particular set of words and each news article is defined by a particular set of topics.

Model

A template formalizing the relationship between an input and an output. Its structure is fixed but it also has parameters that are modifiable; the parameters are adjusted so that the same model with different parameters can be trained on different data to implement different relationships in different tasks.

Natural language processing

Computer methods used to process human language, also called computational linguistics.

Nearest-neighbor methods

Models where we interpret an instance in terms of the most similar training instances. They use the most basic assumption: similar inputs have similar outputs. They are also called instance-, memory-, or case-based methods.

Neural network

A model composed of a network of simple processing units called neurons and connections between neurons called synapses. Each synapse has a direction and a weight, and the weight defines the effect of the neuron before on the neuron after.

Nonparametric methods
Statistical methods that do not make strong assumptions about the properties of the data. Hence they are more flexible, but they may need more data to sufficiently constrain them.

Occam's razor
A philosophical heuristic that advises us to prefer simple explanations to complicated ones.

Online analytical processing (OLAP)
Data analysis software used to extract information from a *data warehouse*. OLAP is user-driven, in the sense that the user thinks of some hypotheses about the process and using OLAP tools checks whether the data supports those hypotheses. Machine learning is more data-driven in the sense that automatic data analysis can find dependencies not previously thought by users.

Outlier detection
An outlier, anomaly, or novelty is an instance that is very different from other instances in the sample. In certain applications such as fraud detection, we are interested in such outliers that are the exceptions to the general rules.

Parallel distributed processing
A computational paradigm where the task is divided into smaller concurrent tasks, each of which can be run on a different processor. By using more processors, the overall computation can be done much faster.

Parametric methods
Statistical methods that make strong assumptions about data. The advantage is that if the assumption holds, they are very efficient in terms of computation and data; the risk is that those assumptions do not always hold.

Pattern recognition
A pattern is a particular configuration of data; for example, 'A' is a composition of three strokes. Pattern recognition is the detection of such patterns.

Perceptron
A perceptron is a type of a *neural network* organized into layers where each layer receives connections from units in the previous layer and feeds its output to the units of the layer that follow.

Population
The set of all possible observable values for a random experiment, a *sample* is a random subset of the population.

Posterior distribution
The distribution of possible values that an unknown parameter can take *after* looking at the data. *Bayes' rule* allows us to combine the *prior distribution* and the data to calculate the posterior distribution.

Precision and recall
Measures used to evaluate an *information retrieval* system. Precision is the ratio of the number of retrieved and relevant records to the number of retrieved records, and recall is the ratio of the number of retrieved and relevant records to the relevant records.

Prior distribution
The distribution of possible values that an unknown parameter can take *before* looking at the data. For example, before estimating the average weight of high school students, we may have a prior belief that it will be between 100 and 200 pounds. Such information is especially useful if we have little data.

Q-learning
A *reinforcement learning* method based on *temporal difference learning*, where the goodness values of actions in states are stored in a table (or function), frequently denoted by Q.

Ranking
This is a task somewhat similar to *regression*, but we care only whether the outputs are in the correct order. For example, for two movies A and B, if the user enjoyed A more than B, we want the score estimate to be higher for A than for B. There are no absolute score values as we have in regression, but only a constraint on their relative values.

Recurrent connection
A type of connection that involves a delay and acts as a short-term memory helping the network to remember its past. The advantage is that the network's output depends not only on its current input but also on the inputs it has seen in previous time steps.

Regression
Estimating a numeric value for a given instance. For example, estimating the price of a used car given the attributes of the car is a regression problem.

Reinforcement learning
It is also known as learning with a critic. The agent takes a sequence of actions and receives a reward/penalty only at the very end, with no feedback during the intermediate actions. Using this limited information, the agent should learn to generate the actions to maximize the reward in later trials. For example, in chess, we do a set of moves, and at the very end, we win or lose the game; so we need to figure out what the actions that led us to this result were and correspondingly credit them.

Sample
A set of observed data. In statistics, we make a difference between a *population* and a sample. Let us say we want to do a study on obesity in high school students. The population is all the high school students, but we cannot possibly observe the weights of all. Instead, we choose a random subset of, for example, 1,000 students and observe their weights. Those 1,000 values are our sample. We analyze the sample to make inferences about the population. Any value we calculate from the sample is a statistic. For example, the average of the weights of the 1,000 students in the sample is a statistic and is an estimator for the mean of the population.

Smart device
A device that has its sensed data represented digitally and is doing some computation on this data. The device may be mobile and it may be online; that is, it may have the ability to exchange data with other smart devices, computers, or the *cloud*.

Speech recognition
Recognizing uttered sentences from acoustic information captured by a microphone.

Supervised learning
A type of machine learning where the model learns to generate the correct output for any input. The model is trained with data prepared by a supervisor who can provide the desired output for a given input. *Classification* and *regression* are examples of supervised learning.

Temporal difference learning

A set of methods for *reinforcement learning* where learning is done by backing up the goodness of the current action to the one that immediately precedes it. An example is the *Q-learning* algorithm.

Transfer learning

Using a model, completely or partially, trained on task A to be used in solving task B. When used in *neural networks*, this corresponds to using some of the layers of the network trained on A also in the network to be trained for B. This can be done if A and B are similar tasks, and is especially useful if we have more data for A than for B.

Validation

Testing the *generalization* performance of a trained model by testing it on data unseen during training. Typically in machine learning, we leave some of our data out as validation data, and after training we test it on this left out data. This validation accuracy is an estimator for how well the model is expected to perform when used later on in real life.

Web scraping

Software that automatically surfs the web and extracts information from web pages.

Preface
1. "Go Master Lee says he quits unable to win over AI Go players." Yonhap News Agency, November 27, 2019. https://en.yna.co.kr/view/AEN2019 1127004800315 (accessed January 29, 2020).

Chapter 1
1. These use the ASCII code devised for the English alphabet and punctuation. The character-encoding schemes we use today cover the different alphabets of different languages.
2. In building portable electronic devices, such as notebook computers, music players, and smartphones, the development of rechargeable lithium-ion batteries was crucial. The 2019 Nobel Prize in chemistry went to three researchers who made this technology possible.
3. It is not the computing power, storage capacity, or connectivity that by themselves produce added value, just as a higher population does not necessarily imply a larger workforce. The enormous number of smartphones in the developing countries does not translate to wealth.
4. A computer program is composed of an algorithm for the task and data structures for the digital representation of the processed information. The title of a seminal book on computer programming is just that: *Algorithms + Data Structures = Programs* (Wirth 1976).
5. Early scientists believed that the existence of rules that explain the physical world is a sign of an ordered universe, which could only be due to a god. Observing nature and trying to fit rules to natural phenomena has an old history, starting in ancient Mesopotamia. Early on, pseudoscience could not be separated from science. In hindsight, the fact that the ancient people believed in astrology is not surprising: If there are regularities and rules about the movement of the sun and the moon, which can be used to predict eclipses for example, positing the existence of regularities and rules about the movement of human beings, which seem so petty in comparison, does not sound far-fetched.

Chapter 2
1. See https://en.wikipedia.org/wiki/Depreciation.

2. Such smoothness assumptions are also frequently used in image processing. For example, when scientists first captured the image of a black hole in 2019, they used the Event Horizon Telescope, which is actually a combination of a number of telescopes around the world each capable of recording only a small part. Smoothness constraints are used to put those pieces together to get the complete image, just like in the illusion of the Kanizsa triangle where we can imagine a complete large triangle even though what we actually see are just small pieces of it.

3. For an excellent history of artificial intelligence, see Nilsson 2009.

4. See Sandel 2012 for some real-life scenarios where decision making based on expected value, or expected utility, may not be the best way. *Pascal's wager* is another example of the application of expected value calculation in an area where it should not be applied.

Chapter 3

1. Here, we are talking about *optical* character recognition where the input is an image; there is also *pen-based* character recognition where the writing is done on a touch-sensitive pad. In such a case, the input is not an image but a sequence of the (x,y) coordinates of the stylus tip, while the character is written on the touch-sensitive surface.

2. Let us say F represents the flu and N represents a runny nose. Using Bayes' rule, we can write the probability that a person has the flu given that we know they have a runny nose:

$$P(F|N) = P(N|F)P(F)/P(N),$$

Here, $P(N|F)$ is the conditional probability in the other direction, namely, that a patient has a runny nose given that they are known to have the flu. $P(F)$ is the probability that a patient has the flu, regardless of whether they have a runny nose or not, and $P(N)$ is the probability that a patient has a runny nose, regardless of whether they have the flu or not.

3. It is interesting that in many science fiction movies, though the robots may be very advanced in terms of vision, speech recognition, and autonomous movement, they still continue to speak in an emotionless, "robotic" voice.

4. This is due to Condorcet's Jury Theorem, which states that in a group that decides by majority voting, if each voter has an independent probability p of voting for the correct decision, for any $p > \frac{1}{2}$ (better than randomly guessing), the probability that the majority vote is correct approaches 1 as the number of voters increase. This is why democracy is better than monarchy (Marquis de

Condorcet supported the French Revolution, though he was later treated as a traitor and sent to prison where he died), but only if the voters can form their own opinions independently, and that is where concepts such as freedom of press and freedom of expression come into play.

Another implication is the advantage of countries that are tolerant to differences, or are open to individuals from different backgrounds, such as immigrants, knowing that differences are a source of diversity and hence of new ideas. Historically speaking, we see that creativity, artistic or scientific, is highest in either countries that were trading nations continuously interacting with different cultures, or countries that were welcoming to immigrants. It is always said that creativity requires being able to "think outside the box"; the more different people are, the less their boxes overlap, and meeting different people enlarges one's box.

5. Bayesian estimation uses Bayes' rule in probability theory (which we saw before) named after Thomas Bayes (1702–1761) who was a Presbyterian minister. The assumption of a prior that exists before and underlies the observable data should have come naturally with the job.

Chapter 4

1. Check thispersondoesnotexist.com for example face images generated by a GAN.

Chapter 7

1. Quoted in "Human Rights and Technology" discussion paper, Australian Human Rights Commission, December 2019, 75, https://www.humanrights .gov.au/our-work/rights-and-freedoms/publications/human-rights-and-technology -discussion-paper-2019.

Chapter 8

1. The speed of light is approximately 300,000 km/sec, so it takes at least 3.33 milliseconds to traverse 1,000 km—distance to a data center. This is not actually such a small number with electronic devices. The connection is never direct and there are always delays due to intermediate routing devices; and remember that to get a response, we need to send a query first, so we need to double the time.

2. For more, see *Frontiers in Massive Data Analysis* (Washington, DC: National Academies Press, 2013).

REFERENCES

Bakshy, E., S. Messing, and L. A. Adamic. 2015. "Exposure to Ideologically Diverse News and Opinion on Facebook." *Science* 348:1130–1132.

Blei, D. 2012. "Probabilistic Topic Models." *Communications of the ACM* 55:77–84.

Bohannon, J. 2015. "Fears of an AI Pioneer." *Science* 349:252.

Börner, K. 2015. *Atlas of Knowledge: Anyone Can Map*. Cambridge, MA: MIT Press.

Brown, N., and T. Sandholm. 2019. "Superhuman AI for Multiplayer Poker." *Science* 365:885–890.

Buchanan, B. G., and E. H. Shortliffe. 1984. *Rule-Based Expert Systems: The MYCIN Experiments of the Stanford Heuristic Programming Project*. Reading, MA: Addison Wesley.

Eisenstein, J. 2019. *Introduction to Natural Language Processing*. Cambridge, MA: MIT Press.

Favaretto, M., E. De Clercq, and B. S. Elger. 2019. "Big Data and Discrimination: Perils, Promises and Solutions: A Systematic Review." *Journal of Big Data* 6 (12): 1–27.

Feldman, J. A., and D. H. Ballard. 1982. "Connectionist Models and Their Properties." *Cognitive Science* 6:205–254.

Fukushima, K. 1980. "Neocognitron: A Self-Organizing Neural Network Model for a Mechanism of Pattern Recognition Unaffected by Shift in Position." *Biological Cybernetics* 36:93–202.

Gilpin, L. H., D. Bau, B. Z. Yuan, A. Bajwa, M. Specter, and L. Kagal. 2018. "Explaining Explanations: An Overview of Interpretation of Machine Learning." ArXiv preprint ArXiv:1806.00069.

Goodfellow, I., Y. Bengio, and A. Courville. 2016. *Deep Learning*. Cambridge, MA: MIT Press.

Goodfellow, I. J. , J. Pouget-Abadie , B. Xu, D. Warde-Farley, S. Ozair, A. Courville, and Y. Bengio. 2014. "Generative Adversarial Networks." ArXiv preprint ArXiv:1406.2661.

Han, J., and M. Kamber. 2011. *Data Mining: Concepts and Techniques.* 3rd ed. San Francisco, CA: Morgan Kaufmann.

Hebb, D. O. 1949. *The Organization of Behavior*. New York: Wiley & Sons.

Hirschberg, J., and C. D. Manning. 2015. "Advances in Natural Language Processing." *Science* 349:261–266.

Hochreiter, S., and J. Schmidhuber. 1997. "Long Short-Term Memory." *Neural Computation* 9:1735–1780.

Horvitz, E., and D. Mulligan. 2015. "Data, Privacy, and the Greater Good." *Science* 349:253–255.

Hubel, D. H. 1995 *Eye, Brain, and Vision*. 2nd ed. New York: W. H. Freeman. http://hubel.med.harvard.edu/index.html.

Jordan, M. I., and T. M. Mitchell. 2015. "Machine Learning: Trends, Perspectives, and Prospects." *Science* 349:255–260.

Koller, D., and N. Friedman. 2009. *Probabilistic Graphical Models.* Cambridge, MA: MIT Press.

Lazic, N., T. Lu, C. Boutilier, M. Ryu, E. Wong, B. Roy, and G. Imwalle. 2018. "Data Center Cooling using Model-Predictive Control." In *Proceedings of the 32nd Conference on Neural Information Processing Systems*, 3818–3827.

LeCun, Y., Y. Bengio, and G. Hinton. 2015. "Deep Learning." *Nature* 521:436–444.

LeCun, Y., B. Boser, J. S. Denker, D. Henderson, R.E. Howard, W. Hubbard, and L. D. Jackel. 1989. "Backpropagation Applied to Handwritten Zip Code Recognition." *Neural Computation* 1:541–551.

Li, Y. 2019. "Reinforcement Learning Applications." ArXiv preprint ArXiv:1908.06973.

Liu, T.-Y. 2011. *Learning to Rank for Information Retrieval*. Heidelberg: Springer.

Marr, D. 1982. *Vision: A Computational Investigation into the Human Representation and Processing of Visual Information*. Cambridge, MA: MIT Press.

McCulloch, W., and W. Pitts. 1943. "A Logical Calculus of the Ideas Immanent in Nervous Activity." *Bulletin of Mathematical Biophysics* 5:115–133.

Mikolov, T., K. Chen, G. Corrado, and J. Dean. 2013. "Efficient Estimation of Word Representations in Vector Space." ArXiv preprint ArXiv:1301.3781.

Minsky, M. L., and S. A. Papert. 1969. *Perceptrons*. Cambridge, MA: MIT Press.

Mitchell, T. 1997. *Machine Learning*. New York: McGraw Hill.

Mnih, V., K. Kavukçuoğlu, D. Silver, A. Rusu, J. Veness, M. G. Bellemare, A. Graves, et al. 2015. "Human-Level Control through Deep Reinforcement Learning." *Nature* 518:529–533.

Moravcik, M., M. Schmid, N. Burch, V. Lisý, D. Morrill, N. Bard, T. Davis, et al. 2017. "DeepStack: Expert-Level Artificial Intelligence in Heads-Up No-Limit Poker." *Science* 356:508–513.

Nilsson, N. J. 2009. *The Quest for Artificial Intelligence: History of Ideas and Achievements.* Cambridge, UK: Cambridge University Press.

Orbanz, P., and Y. W. Teh. 2010. "Bayesian Nonparametric Models." In *Encyclopedia of Machine Learning*. New York: Springer.

Rosenblatt, F. 1962. *Principles of Neurodynamics*. Washington, DC: Spartan Books.

Rumelhart, D. E., G. E. Hinton, and R. J. Williams. 1986. "Learning Representations by Back-Propagating Errors." *Nature* 323:533–536.

Rumelhart, D. E., and J. L. McClelland and the PDP Research Group. 1986. *Parallel Distributed Processing: Explorations in the Microstructure of Cognition*. Cambridge, MA: MIT Press.

Russell, S., and P. Norvig. 2020. *Artificial Intelligence: A Modern Approach*. 4th ed. Hoboken, NJ: Pearson.

Sandel, M. 2012. *What Money Can't Buy: The Moral Limits of Markets*. New York: Farrar, Straus and Giroux.

Schmidhuber, J. 2015. "Deep Learning in Neural Networks: An Overview." *Neural Networks* 61:85–117.

Schwartz, R., J. Dodge, N. A. Smith, and O. Etzioni. 2019. "Green AI." ArXiv preprint ArXiv:1907.10597.

Silver, D., A. Huang, C. J. Maddison, A. Guez. L. Sifre, G. van den Driessche, J. Schrittwieser, et al. 2016. "Mastering the Game of Go with Deep Neural Networks and Tree Search." *Nature* 529:484–489.

Silver, D., T. Hubert, J. Schrittwieser, I Antonoglou, M. Lai, A. Guez, M. Lanctot, et al. 2018. "A General Reinforcement Learning Algorithm that Masters Chess, Shogi, and Go through Self-Play." *Science* 362:1140–1144.

Silver, D., J. Schrittwieser, K. Simonyan, I. Antonoglou, A. Huang, A. Guez, T. Hubert, et al. 2017. "Mastering the Game of Go without Human Knowledge." *Nature* 550:354–359.

Sutton, R. S., and A. G. Barto. 2018. *Reinforcement Learning: An Introduction*. 2nd ed. Cambridge, MA: MIT Press.

Sweeney, L. 2002. "K-Anonymity: A Model for Protecting Privacy." *International Journal of Uncertainty, Fuzziness and Knowledge-Based Systems* 10:557–570.

Tesauro, G. 1995. "Temporal Difference Learning and TD-Gammon." *Communications of the ACM* 38:58–68.

Thrun, S., W. Burgard, and D. Fox. 2005. *Probabilistic Robotics*. Cambridge, MA: MIT Press.

Valiant, L. 1984. "A Theory of the Learnable." *Communications of the ACM* 27:1134–1142.

Vapnik, V. 1998. *Statistical Learning Theory*. New York: Wiley.

Vinyals, O., I. Babuschkin, W. M. Czarnecki, M. Mathieu, A. Dudzik, J. Chung, D. H. Choi, et al. 2019. "Grandmaster Level in StarCraft II Using Multi-Agent Reinforcement Learning." *Nature* 575:350–354.

Vinyals, O., A. Toshev, S. Bengio, and D. Erhan. 2014. "Show and Tell: A Neural Image Caption Generator." Arxiv preprint ArXiv:1411.4555.

Wachter, S., B. Mittelstadt, and C. Russell. 2018. "Counterfactual Explanations without Opening the Black Box: Automated Decisions and the GDPR." *Harvard Journal of Law and Technology,* 31:841–887.

Winston, P. H. 1975. "Learning Structural Descriptions from Examples." In *The Psychology of Computer Vision*, ed. P. H. Winston, 157–209. New York: McGraw-Hill.

Wirth, N. 1976. *Algorithms + Data Structures = Programs*. Upper Saddle River, NJ: Prentice Hall.

Zadeh, L. A. 1965. "Fuzzy Sets." *Information and Control* 8:338–353.

Zoph, B., and Q. V. Le. 2016. "Neural Architecture Search with Reinforcement Learning." ArXiv preprint ArXiv:1611.01578.

FURTHER READING

Duda, R. O., P. E. Hart, and D. G. Stork. 2001. *Pattern Classification*. 2nd ed. New York: Wiley.

Feldman, J. A. 2006. *From Molecule to Metaphor: A Neural Theory of Language*. Cambridge, MA: MIT Press.

Hastie, T., R. Tibshirani, and J. Friedman. 2011. *The Elements of Statistical Learning: Data Mining, Inference, and Prediction*. New York: Springer.

Kohonen, T. 1995. *Self-Organizing Maps*. Berlin: Springer.

Murphy, K. 2012. *Machine Learning: A Probabilistic Perspective*. Cambridge, MA: MIT Press.

Pearl, J. 2000. *Causality: Models, Reasoning, and Inference*. Cambridge, UK: Cambridge University Press.

Witten, I. H., and E. Frank. 2005. *Data Mining: Practical Machine Learning Tools and Techniques*. 2nd ed. San Francisco, CA: Morgan Kaufmann.

INDEX

ETHEM ALPAYDIN is Professor in the Department of Computer Engineering at Özyeğin University, and a member of the Science Academy Society, Istanbul. He is the author of the widely used textbook *Introduction to Machine Learning* (MIT Press), now in its fourth edition.